AI PROOF MANIFESTO

The New Rules of Work in the Age of Intelligent
Machines And the Seven Superpowers to
Rewrite Your Rules

David S. Morgan

TABLE OF CONTENTS

INTRODUCTION

The Great Exposure—Rethinking Everything in the Age of AI

You wake up to an email from HR. Your job is gone—not because you failed, but because AI now does it faster, cheaper, and at scale.

Maybe you saw the warnings. Maybe you didn't. Either way, it's too late.

By 2030, only 33 percent of tasks will be performed solely by humans, compared to 47 percent today (World Economic Forum, 2025). AI isn't "coming for your job"—it's already here, rewriting the rules of work faster than companies or governments can adapt.

Entire professions are being reshaped, compressed, or erased. Roles once considered "safe"—marketing, legal research, financial analysis—are now AI-augmented at best, obsolete at worst.

But here's the truth: AI isn't just eliminating jobs—it's creating new ones, for those who know how to position themselves.

This is not a distant future. It's happening now. And waiting is not a strategy.

I Thought I Had Figured It Out. Then AI Exposed Me.

In the late spring of 2024, I was finalizing my book, THRIVE: A Self-Starter's Guide to Accelerating Your Career—a playbook built on adaptability, skill-building, and strategic reinvention.

I had poured years into crafting it. Late nights. Decades of experience distilled into a blueprint for future-proofing careers.

And then, as I reviewed the final manuscript, a realization hit me like a freight train:

AI had already disrupted everything I had written. But disruption isn't the enemy—inaction is.

The career strategies I once believed were timeless were already crumbling.

If AI could make my entire playbook irrelevant overnight, what would it reveal about yours?

It was like standing on a bridge, urging others to cross—only to watch it collapse beneath me.

AI Isn't Replacing You—It's Exposing You

For decades, we were sold a promise about work:

- Get a degree.
- Work hard.
- Climb the corporate ladder.
- And you'd be "safe."

They never told us AI would burn that ladder to the ground before we reached the top.

AI doesn't just automate tasks—it exposes the gap between those who think and those who execute. It forces a reckoning: What makes you indispensable? What do you offer that AI can't replicate?

Because the only real security left isn't in what you know, but in how you think, adapt, and create.

Those who own their human edge—curiosity, adaptability, creativity—will dominate. Those who don't will fade into irrelevance.

Exposed. Employees coasting on predictable tasks? Exposed. Exposed. Managers who don't actually manage? Exposed. Exposed. Professionals whose expertise is now just an AI prompt away? Exposed.

If your work is repeatable, it's already being automated. If your value is execution, it's already commoditized. If you're waiting for direction, you're already behind. But if you're willing to rethink, reposition, and retool—AI becomes your advantage, not your adversary.

This isn't a warning. It's a reality check.

In 2024, U.S.-based tech firms cut over 95,000 jobs, many in roles AI could automate overnight—customer support, marketing, legal research (Robson, 2025).

Marketing teams that once spent days crafting campaigns now watch AI do it in seconds.

Paralegals who once pored over case law now compete with AI that delivers instant insights.

Even doctors—once considered untouchable—are seeing AI outperform human accuracy in diagnostics.

AI doesn't care how hard a job is. It cares how scalable it is.

The Hard Truth: AI Doesn't Care About Your Degree—But It Rewards Those Who Adapt

I've lived through disruption before.

I've zigzagged across industries—manufacturing, tech, nonprofits, education. I've reinvented myself multiple times and even patented assistive technology for the blind.

Each shift forced me to rethink everything. But one truth remains:

AI will never replace:

- Curiosity—to ask better questions. That tiny moment—a spark, a twitch—when something clicks and you start to wonder. AI doesn't have that impulse. But you do.
- Synthesis—to connect ideas in ways AI can't. AI retrieves. You explore. AI recombines. You invent.
- Initiative—to act before waiting for permission. AI reacts. You decide.

These aren't just skills. They're your seven superpowers —the human edge that AI can't replace. And they're the key to winning in the AI era.

When I pivoted industries, curiosity drove me to learn fast.

When I patented new technology, synthesis let me combine insights from multiple fields.

When AI disrupted my last book, initiative pushed me to act before the dust settled.

That's how I AI-proofed myself.

And that's why I'm still here.

The real risk isn't that AI will take your job.

The real risk is assuming you have no control over what happens next.

THE THREE CHOICES IN THE AI ERA

Right now, you stand at a crossroads.

- Ignore AI, and it will ignore you—until it replaces you.

- Use AI, and you might work faster—but you'll still lose to those who think bigger.
- AI-proof yourself, and you will reinvent your value before AI does it for you.

The path you choose today will determine whether AI becomes your greatest threat or your most powerful advantage.

Where We're Headed

This book is not a technology manual.

It is a career revolution.

Here's how we'll AI-proof your future:

- Part 1: The AI Lies You've Been Told—Dismantle the myths that hold you back. AI doesn't steal jobs; it exposes predictability.
- Part 2: How to Become an AI Architect—Rewire your mindset. Stop waiting like an employee; start designing like a visionary.
- Part 3: Your AI-Proof Playbook—Execute now. Master the uniquely human superpowers that AI can't replicate—and use them to shape your future before AI shapes it for you.

Your Choice Starts Now

AI isn't waiting.

The world is splitting:

- Those who command AI.
- Those who are commanded by it.

I wrote this book not to explain AI, but to show you how to make it work for you—so that you can shape your future on your terms.

The choice is yours. Let's get started.

Next, in Chapter 1, we dismantle the biggest threat you face—not AI itself, but the dangerous career-ending myths that keep you from adapting.

Because the greatest risk right now isn't AI.

It's believing the wrong things about it.

PART 1

THE AI LIES YOU'VE BEEN TOLD

The biggest threat isn't AI—it's believing the wrong things about it. This section exposes the myths holding you back and resets your mindset for the AI era.

WHERE WE'RE HEADED

The biggest threat isn't AI—it's believing the wrong things about it. In this section, we dismantle the myths that are keeping people stuck. You'll learn why AI isn't replacing jobs the way the media suggests, how automation exposes stagnation rather than causes it, and why the real danger isn't AI itself—but failing to adapt to it. Before we move into strategy and execution, we need to break free from outdated assumptions about work, value, and career security.

AI Isn't Replacing You—It's Exposing You

AI is a mirror, not a monster—if your work is predictable, you're already obsolete.

But exposure isn't the same as elimination. AI reveals gaps—but it also creates new openings. The shift we're experiencing isn't about making human work obsolete—it's about forcing us to rethink where our true value lies. And that's something you can control. The right strategies will help you position yourself for what's next.

This book isn't about which jobs AI will take. It's about the truth AI is already revealing.

Building on what we've established in the Introduction, the real disruption isn't that artificial intelligence is making humans obsolete—**AI isn't creating your irrelevance, it's accelerating the revelation of the irrelevance that was always there**.

The real threat isn't ChatGPT or Midjourney or whatever new model dropped last week. The real threat is continuing to work like it's 2019.

Let's be brutally honest: If your job primarily involves following patterns you've already mastered, you're not just at risk—you're already obsolete. You just don't know it yet.

THE GREAT EXPOSURE IN ACTION

The day Marcus lost his job wasn't the day ChatGPT got smarter—it was the day he realized how **predictable** his work had become.

For six years, Marcus was the financial analyst everyone respected at Meridian Capital. His PowerPoint decks? Legendary. His Excel models? Elegant. His quarterly reports on Latin American markets? Comprehensive. He spoke Spanish, maintained connections in Mexico City and São Paulo, and had developed models tracking economic indicators across the region.

"I was the go-to guy for Latin American markets," Marcus told me. "My connections and cultural understanding gave me insights no one else had. I thought my work was too nuanced, too relationship-based for a machine to replicate."

Then came that Tuesday morning. His manager called him into the glass-walled conference room and slid a tablet across the table.

"This is what the AI produced overnight," she said, pointing to a financial analysis eerily similar to the one Marcus had spent three days creating the previous week. "And this," she continued, sliding another document forward, "is the analysis you delivered last quarter."

4

The similarities weren't just close. They were **identical** in insight, if not in style.

"We're restructuring the department," his manager explained, not quite meeting his eyes. "We're keeping analysts who can work **above** what the AI delivers."

As Marcus packed his desk that afternoon, the truth hit him with brutal clarity: AI hadn't replaced him. It had **exposed** him.

"What really shook me," Marcus later told me, "was when my boss pulled up five of my previous reports and showed how they followed almost identical structures, used similar phrases, and even made comparable recommendations when facing similar market conditions. I had been telling myself my work was unique and irreplaceable, but seeing it side-by-side with the AI output was a wake-up call. About 80% of what I did could be automated."

His reaction followed a familiar pattern: denial, resistance, then panic.

"I started working longer hours, adding more detail to my reports, trying to prove my value. But I was competing on the wrong metrics. I was trying to out-detail the AI, which is like trying to outrun a Ferrari."

The critical mistake Marcus made was identifying with the tasks rather than the outcomes. He saw himself as a "report creator" rather than an "investment insight provider." This task-based identity made him vulnerable when AI could perform the same tasks more efficiently.

"The most painful realization was that much of my so-called experience was just repetition," Marcus admitted. "I had been doing essentially the same analyses for years, just with different data. I hadn't substantially evolved my approach in nearly a decade."

Six months after the AI implementation, Marcus was let go during a department restructuring.

Six months later, Marcus sat in a coffee shop, scrolling through job postings with a sinking feeling. The roles he once considered within reach now listed requirements that made him pause—"AI proficiency preferred," "strong experience in automation tools." Even positions he felt qualified for received dozens of applications within hours. He had applied to 47 jobs. Four interviews. No offers.

One evening, a former colleague, James, reached out. Over drinks, James shared how he had pivoted. "When the layoffs hit, I knew I had to learn fast. I dove into AI-assisted financial modeling, started a Substack on emerging markets, and leveraged AI tools to offer consulting. Now I contract with three firms. It's not about being the best analyst anymore—it's about being the one who sees what's coming."

Marcus clenched his jaw. He had spent months waiting for an opportunity that no longer existed. James hadn't waited. He adapted. That realization cut deeper than the layoff itself.

Down the hall, Elena Morales faced the same AI disruption with a radically different outcome. With eight years analyzing Asian markets at Meridian, her response wasn't resistance—it was curiosity.

"I asked to be part of the pilot program," she told me. "I wanted to understand exactly what this technology could and couldn't do."

Within weeks, Elena had mapped the AI's capabilities against her workflow. "I realized it could handle about 70% of my technical analysis tasks—gathering data, running models, creating visualizations, drafting preliminary findings. But there were clear limitations."

Instead of competing with the AI on its strengths, Elena pivoted to focus on its weaknesses. "I recognized that AI would handle the 'what' while humans would own the 'why' and 'so what,'" she explained.

Elena restructured her role around three areas where the AI struggled:

1. Contextualizing data within political and cultural developments not captured in the numbers
2. Building and maintaining relationships with local sources across Asian markets
3. Translating findings into strategic recommendations tailored to specific client needs and risk profiles

"Instead of seeing AI as a threat, I positioned it as a tool that eliminated the least interesting parts of my work," Elena said. "I became known as someone who could make AI outputs more valuable by adding the human layer of analysis."

Within eight months of the AI implementation, Elena was promoted to Senior Strategic Advisor with a 40% increase in compensation. Today, she leads a team of analysts who all use AI as a force multiplier rather than viewing it as competition.

"The analysts who thrived weren't necessarily the most technically skilled," noted James Harmon, Meridian's CTO. "They were the ones who quickly redefined their value in relation to what the AI could do, rather than in opposition to it."

WHAT'S DIFFERENT THIS TIME: THE VELOCITY OF EXPOSURE

We're not witnessing a great replacement. We're witnessing a **great exposure.**

AI functions as a mirror, reflecting the parts of our work that were always formula-driven, pattern-based, and ultimately low-value. It's revealing which aspects of our "expertise" were actually just memorization and pattern recognition.

This isn't new. Technology has always exposed human limitations:

- Calculators exposed that mental arithmetic wasn't the essence of mathematics
- Word processors exposed that neat handwriting wasn't the core of good writing
- The internet exposed that memorizing facts wasn't the heart of intelligence

What's different this time isn't the nature of the change—it's the velocity.

The exposure that used to unfold over decades now happens **almost overnight**. The skills that once took years to build can become **commoditized within months—sometimes weeks.**

Consider this: Recent research from McKinsey suggests that approximately **60% of tasks in white-collar professions** follow predictable enough patterns to be handled by current **generative AI systems** (McKinsey Global Institute, 2023). Not future AI. **Today's AI.**

This rapid acceleration aligns with **Brynjolfsson's latest findings**, which emphasize that AI is not merely reshaping the workforce—

it is **exposing and accelerating** task automation at an unprecedented rate. In a **2025 World Economic Forum discussion**, Brynjolfsson noted that while **many companies are aggressively investing in AI**, they are struggling to realize productivity gains because they are still in an **experimental phase** rather than fully integrating AI into high-impact tasks. He argues that **AI's true disruption isn't just in replacing jobs, but in revealing which work is inherently routine and susceptible to automation** (Brynjolfsson, 2025).

This supports the **"great exposure" thesis**, which asserts that AI acts as a **mirror**—revealing the **formulaic versus the truly creative** aspects of work at an unprecedented speed. Organizations and professionals who fail to adapt **will not have the luxury of time**; the shift from valuable skill to commodity **is no longer measured in decades but in quarters, sometimes even weeks**.

Let that sink in.

"But my job requires creativity/expertise/judgment," you protest.

Maybe. But **how much of that creativity follows formulas you don't even recognize?** How much of that expertise consists of patterns you've internalized? How much of that judgment follows decision trees you could—if pressed—write down?

AI doesn't create your irrelevance. It simply reveals it faster than you can hide it.

WARNING SIGNS OF EXPOSURE

Most people won't recognize their exposure until it's too late. Here's how to diagnose your risk today. Watch for these warning signals:

1. **Your deliverables follow predictable structures -** Reports, analyses, or creative work that follow consistent templates or approaches

1. **You've been doing essentially the same work for years -** Just with different variables or contexts
2. **You spend most of your time on execution rather than strategy -** Implementing rather than defining what should be implemented
3. **Colleagues ask fewer "how" questions and more "why" questions** about your work - Indicating the process is becoming less mysterious
4. **You find yourself defending traditional approaches -** Rather than exploring new possibilities
5. **Your work products can be evaluated by clear metrics** - The more objectively your work can be measured, the more likely it can be automated
6. **You rarely face truly novel problems -** If most challenges you encounter are variations of familiar situations, you're highly exposed

Marcus missed all these signals. Elena spotted them immediately and acted. The difference wasn't just awareness—it was a willingness to confront uncomfortable truths about professional value.

THE VALUE HIERARCHY HAS BEEN INVERTED

Elena succeeded because she mastered what AI couldn't. The real question is: What skills will define success in the AI era?

For decades, we've operated under a relatively stable professional value hierarchy:

- Technical skills at the top
- Analytical abilities next
- Communication abilities after that
- "Soft skills" grudgingly acknowledged somewhere near the bottom

AI IS RAPIDLY INVERTING THIS PYRAMID.

Skills that commanded premium salaries just months ago are being commoditized overnight:

- Basic data analysis: **Commoditized**
- First-draft writing: **Commoditized**
- Routine coding: **Commoditized**
- Market research compilation: **Commoditized**
- Design based on existing patterns: **Commoditized**

Meanwhile, previously undervalued skills are suddenly commanding a premium:

- Critical questioning
- Contextual judgment
- Taste and discernment
- Genuine empathy and relationship-building
- Original thinking (not just creative execution)

Take Sarah, a marketing director who built her career on being "the creative one." When her team started using AI to generate campaign concepts, she initially scoffed. "A machine can't replicate my creativity," she insisted.

Then she saw the AI's output. Not only could it generate dozens of concepts in minutes, but some were arguably more original than what her team typically produced.

The uncomfortable realization: much of what she considered "creativity" was actually pattern recognition and recombination—skills that AI excels at.

The painful truth wasn't that AI was creative. It was that her creativity had become formulaic without her noticing. **AI didn't steal her creative identity; it exposed limitations she hadn't recognized.**

According to research from Harvard Business School, when professionals are asked to evaluate their own distinctiveness, they consistently overestimate it by 20-30% (Gino, 2018). We think our approaches are unique when they often follow predictable patterns that algorithms can easily detect and replicate.

THE SPEED GAP: THE NEW DIGITAL DIVIDE

The most dangerous gap emerging in professional life isn't between the technically skilled and unskilled. It's between fast adapters and slow adapters—what I call the **adaptation velocity gap**.

Those who quickly adapt to AI systems are creating an advantage that widens exponentially, not linearly. While traditional skill gaps could be closed with education and experience, adaptation velocity gaps compound over time, becoming nearly unbridgeable.

Think about it this way: If you and a competitor both start businesses today, and they use AI to operate 20% more efficiently than you, after one year they'll be slightly ahead. But that additional margin allows them to innovate and adapt even faster, widening the gap to perhaps 30% by year two. By year three, they might be operating at twice your efficiency, investing in areas you can't afford to explore, and capturing opportunities you can't even see.

This isn't theoretical.

AI Productivity Impact *Early adopters of AI coding assistants report significant productivity gains. Developers using tools like GitHub Copilot are completing more pull requests weekly, with newer developers often seeing the largest productivity boosts* (GitHub, 2023).

Tool Adoption Trends *Companies integrating AI into development workflows report that teams can complete certain tasks significantly faster than before. Engineers are spending less time on repetitive coding and more time on creative problem-solving* (Stack Overflow, 2023).

Usage Patterns *Analysis of how developers interact with AI coding tools reveals that many regularly accept AI-generated code suggestions, particularly for routine coding tasks. This allows them to focus on more complex aspects of software development* (GitHub, 2023).

Firms that effectively integrate advanced digital tools, like generative AI, don't just edge out competitors—they unlock accelerating productivity gains over time. A 2023 study from the Stanford Digital Economy Lab showed that a Fortune 500 company using AI in customer service boosted issue resolution rates by 14% on average, with novice workers seeing gains up to 34% as they adapted, widening performance gaps with less adaptive peers (Brynjolfsson, Li, & Raymond, 2023)

While you're deciding whether to learn how to use AI, others are already using it to reinvent entire business models. By the time you get comfortable with today's AI capabilities, they'll be mastering tomorrow's innovations and widening the competitive gap even further.

The real impact of this velocity gap? As AI adoption accelerates, many middle-management positions may be eliminated or radically transformed due to these adoption disparities. The divide isn't just about who uses AI—it's about how quickly they leverage it to create new value paradigms.

According to the Summer 2023 Fortune/Deloitte CEO Survey, 83% of CEOs indicated they are likely to adjust workforce skills and training over the next six months to keep pace with rapid advancements in AI and emerging technologies (Deloitte Insights & Fortune, 2023). This shift is already underway, with nearly half also considering changes to work arrangements—like hybrid or remote setups—based on technology's evolving role in their operations.

This isn't a future concern—it's happening now.

PwC's 2025 AI Business Predictions report highlights that AI is already transforming industries, with leaders integrating AI agents into workforces to double knowledge work capacity and boost productivity by up to 50%. This isn't a distant vision—companies are actively redefining roles and strategies now to stay competitive in an AI-driven economy (PwC, 2024).

The Exposed vs. The Adapted: A Comparison

Dimension	Exposed Professionals	Adapted Professionals
Focus	Tasks (creating reports, writing code)	Outcomes (improving decisions, solving problems)
Response to AI	Resistance or over-detailing	Leverage and strategic elevation
Adaptation Speed	Slow, incremental change	Rapid, transformative evolution
Value Source	Efficiency in routine work	Judgment in ambiguous situations
Primary Identity	Role-based ("I'm an analyst")	Impact-based ("I improve investment decisions")
Career Trajectory	Diminishing returns	Expanding opportunities
Time Allocation	Mostly execution	Balance of strategy and execution
Professional Growth	Linear, within domain	Exponential, across domains

THE NEW METRICS OF VALUE

Experience is Dead Currency

For generations, years of experience has been among the most valued credentials in professional life. Job listings demanded it. Resumes highlighted it. Compensation was tied to it.

15

In the AI era, experience is rapidly being devalued—not because it's worthless, but because its meaning has fundamentally changed.

The problem isn't just that AI can instantly access more experience (in the form of data) than any human could accumulate in a lifetime. It's that **AI is collapsing experience curves across professions**. Tasks that once required 10,000 hours of human practice can now be performed adequately by someone with 1,000 hours of experience plus AI assistance.

Consider legal research. Traditionally, junior associates spent years learning how to efficiently navigate case law, developing pattern recognition that allowed them to quickly identify relevant precedents. Today, legal AI systems can search millions of cases in seconds, identifying relevant precedents with accuracy that rivals experienced attorneys.

This doesn't make legal experience irrelevant—but it dramatically changes which aspects of that experience remain valuable. The ability to construct novel legal arguments, understand client needs, and navigate complex stakeholder relationships becomes more important than research efficiency or precedent recognition.

As AI systems continue to evolve, much of the tacit knowledge currently held exclusively by senior professionals will become more widely accessible. This democratization of expertise will permanently alter how experience is valued in the market.

Across fields, we're seeing the same pattern: **the experience that matters isn't time served, but adaptability demonstrated**. Your adaptability quotient now matters more than your intelligence quotient.

The Rise of Decidedly Human Metrics

As traditional experience decreases in value, new metrics are emerging to measure professional worth:

2. **Judgment in ambiguity**: How well do you make decisions with incomplete information and unclear outcomes? While AI excels with defined parameters, human judgment in novel situations remains superior.

3. **Question quality**: The ability to ask incisive questions that reframe problems is becoming more valuable than having ready answers. AI can answer questions, but it struggles to identify which questions should be asked.

4. **Novel pattern recognition**: Identifying patterns across seemingly unrelated domains—combining insights from biology, psychology, and economics to solve a business problem, for instance—remains a uniquely human strength.

5. **Human-to-human influence**: The capacity to build trust, navigate office politics, and inspire others cannot be automated. Leadership that motivates human action is immune to AI replacement.

6. **Values-based decision making**: Determinations that require weighing ethical considerations, company values, and long-term brand implications need human judgment informed by shared cultural context.

Forward-thinking organizations are evolving their performance review processes to prioritize adaptability, collaboration, and learning agility alongside traditional productivity metrics. Many leading companies are now measuring "learning velocity"—the ability to rapidly acquire and apply new skills—as a key performance indicator.

THE EXPANSION-CONTRACTION PARADOX

As AI capabilities expand, something counterintuitive is happening to the human value zone—it's simultaneously contracting and expanding.

The contraction is obvious: Routine aspects of knowledge work are rapidly losing value. Tasks that follow clear patterns or can be reduced to rules are being automated at breathtaking speed.

Less obvious is the expansion: New value territories are opening up precisely because of AI capabilities. These expanded zones require uniquely human abilities that work in concert with AI rather than in competition with it.

Consider financial advising. AI is rapidly commoditizing basic investment advice and portfolio construction—contracting the traditional value zone for financial advisors. Simultaneously, it's expanding opportunities for advisors who can address the psychological and emotional aspects of money management, coach clients through complex life transitions, and build trust during market volatility.

In the coming years, many of today's job descriptions will be rendered obsolete, while entirely new roles centered on human-AI collaboration will emerge. This isn't just job displacement—it's a fundamental realignment of human value in the workplace.

The strategic imperative is clear: **Position yourself in the expansion zones, not the contraction zones.** The professionals who thrive won't be those who resist AI encroachment on contracting territory, but those who stake early claims on expanding territory.

BECOMING EXPOSURE-PROOF: YOUR AI VULNERABILITY AUDIT

Before you can become exposure-proof, you need to honestly assess your vulnerability. Consider this your personal AI exposure audit—a framework for evaluating which aspects of your work are most likely to be revealed as low-value by advancing AI.

Score yourself from 1-10 on each question below, where 1 represents high vulnerability and 10 represents high resilience.

1. Pattern Predictability Score (___/10)

What percentage of my work follows predictable patterns? Be brutally honest. If you've been doing your job for more than two years, you've likely developed routines and approaches that you repeat. What percentage of your weekly tasks are essentially variations of the same work with different variables?

- **Score 1-3**: Over 70% of your work follows predictable patterns
- **Score 4-6**: 40-70% of your work follows predictable patterns
- **Score 7-10**: Less than 40% of your work follows predictable patterns

2. Value Source Score (___/10)

How much of my value comes from efficiency versus insight? If your professional worth is primarily tied to doing known tasks quickly and accurately, you're highly exposed. If your value comes from generating novel insights or making judgment calls that can't be reduced to rules, you're more insulated.

- **Score 1-3**: Most value comes from efficiency and accuracy in known tasks
- **Score 4-6**: Value comes from both efficiency and occasional insights
- **Score 7-10**: Most value comes from novel insights and judgment calls

3. Automation Resilience Score (__/10)

What would happen if everything I do that can be documented could be done instantly by AI? This thought experiment separates the truly exposure-proof professionals from the temporarily safe. Imagine everything you do that could be written down as a process suddenly costs nothing and takes no time. What unique value would you still provide?

- **Score 1-3**: Limited or no unique value remains
- **Score 4-6**: Some unique value remains but significantly diminished
- **Score 7-10**: Core value proposition remains intact

Total AI Exposure Score: __/30

- **20-30**: Relatively Exposure-Proof -- Focus on leveraging AI to enhance your already strong position
- **10-19**: Moderately Exposed -- Need strategic repositioning within your field
- **0-9**: Highly Exposed -- Require immediate and significant reinvention

FROM TASK-DOER TO OUTCOME-OWNER

The most crucial mindset shift in becoming exposure-proof is moving from identifying with tasks to identifying with outcomes.

Task-doers say: "I create financial models" or "I write marketing copy" or "I manage client accounts."

Outcome-owners say: "I improve investment decision quality" or "I drive customer acquisition" or "I build client loyalty."

This shift isn't just semantic—it's strategic. Tasks can be automated. Outcomes require orchestration across multiple capabilities, including but not limited to AI.

This transition is psychologically difficult because most of us have built professional identities around mastering specific tasks. We take pride in being excellent writers, skilled analysts, or creative designers. Letting go of task mastery as our core identity feels like professional death.

But here's the liberating truth: **You were never actually the tasks you performed. You were always the outcomes you enabled.** The tasks were merely the means available at the time.

To redefine your role around outcomes:

1. **Identify the fundamental business need you serve** What problem does your work ultimately solve? What opportunity does it capture? What risk does it mitigate?
2. **Map the complete outcome journey** What sequence of events leads to the outcome being achieved? Where are you currently involved? Where could you add value beyond your current involvement?

3. **Divorce strategies from tactics** Strategies define the "why" and "what" of achieving outcomes. Tactics are the "how." As AI increasingly handles tactics, your value lies in owning the strategy.

Remember: **The person who defines the problem will always be more valuable than the one who executes the solution.** As AI gets better at execution, the premium on problem definition grows exponentially.

Leveraging the Exposure to Your Advantage

Once you understand your exposure and have shifted to an outcome orientation, you can choose from four strategic paths to thrive in the AI era:

1. **ELEVATE**: Move up to more complex judgment work
   o Shift your focus to problems with higher ambiguity, greater stakeholder complexity, and more significant business impact
   o These areas require contextual understanding and sophisticated judgment that AI can support but not replace
   o *Example*: A financial analyst who transitions from producing routine market reports to synthesizing multi-domain insights for strategic investment decisions
2. **INTEGRATE**: Become the human-AI collaboration specialist
   o Position yourself as the expert in optimally combining human and AI capabilities

- o Develop deep knowledge of AI strengths and limitations in your domain
- o Become the orchestrator who designs workflows that leverage both
- o *Example*: A marketing professional who becomes the specialist in designing campaigns that combine AI-generated creative variations with human-directed strategic positioning

3. **TRANSLATE**: Position yourself at the interface between AI outputs and human needs
   - o Focus on making AI outputs useful, relevant, and trusted by humans
   - o Develop skills in contextualizing, customizing, and communicating AI-generated insights in ways that drive human action
   - o *Example*: A data scientist who shifts from building models to translating analytical outputs into strategic recommendations for executives who don't trust "black box" solutions

4. **SPECIALIZE**: Deepen expertise in areas AI struggles with
   - o Identify aspects of your field that require embodied experience, cultural context, emotional intelligence, or ethical judgment
   - o Become the go-to specialist for these dimensions
   - o *Example*: A customer service manager who builds expertise in handling emotionally complex customer situations that require nuanced human empathy

The goal isn't to outrun AI—it's to run where AI can't yet follow, while simultaneously harnessing AI's power to make your unique human capabilities more valuable.

Your Wake-Up Call

AI isn't replacing humans—it's replacing specific modes of working that have dominated professional life for decades. The wave of AI we're experiencing now isn't just another technology shift; it's a fundamental reorganization of what creates value in knowledge work.

This leaves you with a stark choice: **Be exposed by AI or be elevated by it.**

The uncomfortable truth is that this disruption won't affect everyone equally. It will create unprecedented opportunities for those willing to shed old paradigms while leaving behind those who cling to fading notions of professional value.

As AI transforms the workplace, displacing some jobs while creating others, enterprises are rapidly adopting the technology. Gartner research shows that by 2026, over 80% of enterprises will have used generative AI APIs or deployed AI-enabled applications, up from less than 5% in 2023 (Gartner, 2023). This surge signals significant changes to work processes in the near term.

The question isn't whether change is coming—it's whether you'll be among those displaced or those who seize new opportunities.

The most dangerous response is the most common one: acknowledging AI's importance while making only incremental adjustments to how you work. This middle path leads straight to irrelevance.

Instead, embrace radical reinvention. Stop competing with AI. Start designing a future where AI makes you exponentially more valuable by handling everything you shouldn't be doing anyway.

ACTION BOX: TAKING YOUR FIRST STEP

Don't just read this—act on it.

Take 10 minutes right now to complete your AI Vulnerability Audit above. Be brutally honest with yourself. If your score is 6 or below in any category, you're in the danger zone and need immediate action.

Now identify ONE task you currently perform that could be handled by AI, freeing you to focus on higher-value work. This single shift is your first step toward becoming AI-proof.

The mirror of AI is already reflecting your professional reality. Will you look away, or will you use what you see to transform yourself?

Now that you understand the real threat isn't AI but adaptation speed, let's explore why traditional career paths are becoming dead ends in the age of AI—and what will replace them.

THREE THINGS FOR THIS WEEK

Reality Check: AI Isn't Coming for Your Job—Your Complacency Is

1. Identify one aspect of your job that AI could automate, then brainstorm how you can elevate your role beyond execution.
2. Spend 30 minutes exploring an AI tool relevant to your field—what can it do, and what are its limits?
3. Start a **Pattern Awareness Journal**—each day, note one repetitive task you do that AI could take over.

02

THE CAREER LADDER IS DEAD AND AI BURNED THE RUNGS

Your safe, linear career path is gone. Time to unlearn old career rules.

When Robert first integrated an AI system into his architectural firm, he expected to cut costs and reduce headcount. Six months later, he had hired **three additional architects**.

"It makes no sense on paper," he told me. "We automated 40% of our design process. So why do I need **more** architects, not fewer?"

The answer came when Robert's firm won the biggest project in their 15-year history—a sustainable urban development that would have been beyond their capabilities before.

"The AI handled the repetitive tasks—the structural calculations, the material specifications, the code compliance checks. This freed my team to push boundaries on the creative and human-centered aspects of design. We delivered a proposal that was both technically flawless and breathtakingly innovative. Our client said no other firm managed both."

After implementing AI, the firm's output increased by 60%, allowing them to secure three times more high-profile projects than the previous year. Their revenue grew by 85% while their design quality—measured by sustainability metrics and client satisfaction scores—improved by 40%.

Across town, another architectural firm faced a different outcome. James, the founder, dismissed AI tools as "expensive toys" that would "homogenize design" and "undermine the creative soul of architecture." While Robert's firm was expanding, James's was shrinking—losing contracts to competitors who delivered comparable designs in half the time at lower costs. Within 18 months, James had laid off 30% of his staff and was considering selling the business.

This contrast illustrates the reality of the AI revolution: **AI doesn't replace people. It replaces predictable work. Your job? Stay unpredictable.** AI isn't eliminating jobs-it's eliminating professionals who cant evolve fast enough.

The future doesn't belong to those who resist AI. Nor does it belong to those who merely adapt to AI. It belongs to those who become **Augmented Professionals**—people who understand exactly how to leverage AI strengths while doubling down on uniquely human capabilities.

The End of Professional Development as We Know It

For decades, career advancement followed a predictable pattern: master technical skills, accumulate experience, climb the organizational ladder. That model is now fundamentally broken. **In the AI era, the traditional career ladder isn't just changing—it's been completely dismantled.**

The comfortable "skills escalator" that carried professionals upward through their careers has stopped working. No longer can you learn a set of skills, practice them for years, and expect to remain valuable. The half-life of professional skills has collapsed from decades to months.

While AI is undeniably compressing or eliminating roles, history shows that every major technological disruption doesn't just destroy—it reinvents. The key isn't whether jobs disappear, but whether new opportunities emerge fast enough to replace them. AI is no exception.

By 2030, 39% of workers' core skills are expected to change, according to the World Economic Forum's *Future of Jobs Report 2025* (World Economic Forum, 2025). Technical skills alone won't future-proof careers—what matters is adaptability, resilience, and the ability to integrate AI into high-impact decision-making. The professionals who cling to old expertise will find themselves trapped in skill obsolescence cycles, while those who continuously retool will command new opportunities before others even realize the landscape has changed.

What's more shocking is that **the very idea of climbing upward is becoming obsolete.** Career progression isn't vertical anymore—it's multidimensional. It's not about climbing higher; it's about expanding your capabilities across domains that AI cannot touch.

Traditional job titles—those comforting labels that defined professional identity for generations—are rapidly losing meaning. When AI can perform many of the tasks that once defined a role, what exactly does that title represent anymore?

Consider this: The marketing director who can't effectively partner with AI is now less valuable than the marketing coordinator who can. The senior developer who resists AI coding tools is being outperformed by the junior developer who embraces them. **The hierarchy has been inverted, and the rungs aren't just burned—-they're gone.**

THE AI LADDER PARADOX

While it's tempting to see AI as purely destructive, history tells a different story. Every great technological shift hasn't just destroyed jobs—it has created new ones that never existed before. AI is no exception.

While AI is eliminating traditional career steps, it's simultaneously creating new, higher-level opportunities that weren't previously accessible. This creates an intriguing paradox: the same force destroying familiar career paths is generating entirely new ones.

Think of what happened during the Industrial Revolution. As automated looms replaced handweavers, they eliminated an entire career progression—from apprentice to journeyman to master weaver. Yet simultaneously, they created entirely new career paths in machine operation, mechanical engineering, factory management, and textile design. The economic value didn't disappear—it transformed and redistributed.

We're seeing the same pattern today but at an accelerated pace. AI is compressing or eliminating entry and mid-level positions in many fields while creating specialized opportunities at higher levels.

Most professionals assume AI adoption is driven from the top down, but the opposite is true. A McKinsey survey found that three times more employees are actively using AI for at least 30% of their work than their leaders assume (Mayer et al., 2025). This means that the true winners in the AI era aren't those waiting for permission from management but those who self-initiate, experiment, and integrate AI into their workflow before it becomes mandatory.

The most significant insight isn't just that AI eliminates some jobs while creating others. It's that AI fundamentally changes how careers develop. Instead of a predictable ladder with clearly defined steps, professionals now navigate a complex network of opportunities where advancement depends less on time served and more on the unique value combinations they create.

This raises a critical question: If traditional advancement paths are disappearing in your industry, what new "rungs" are appearing that you might climb? The greatest opportunities often emerge at the intersections—where human judgment meets machine capability, where domain expertise meets technological integration, where established practices meet innovative approaches.

A Third Path: The Cross-Functional Pivot

Between Robert's architectural firm expansion and James's decline, there's a critical middle path that many professionals will navigate. Consider Elena, a mid-level marketing manager at a consumer goods company.

When generative AI tools first appeared, Elena neither dismissed them nor immediately championed them. Instead, she took a measured approach: "I needed to understand how these tools might change our work before deciding how to respond."

Elena began by identifying which aspects of her team's work involved pattern recognition and rule-based execution—tasks where AI excelled—versus those requiring contextual judgment and creative thinking. She discovered that about 40% of her team's work fell into the first category.

"My first instinct was to protect those tasks—to argue that humans still did them better. But that wasn't true, and I knew it wouldn't be a winning strategy long-term."

Instead, Elena proposed a three-month experiment. Her team would use AI tools to handle content optimization, basic market analysis, and campaign performance tracking, while refocusing their human energy on consumer psychology, brand storytelling, and strategic innovation.

The results were revealing: overall productivity increased by 35%, but more importantly, the team developed entirely new capabilities at the intersection of data science and creative strategy. They created a novel approach to "narrative-driven personalization" that combined AI-powered customization with human-crafted emotional storytelling.

When the company restructured its marketing division, traditional department lines were erased. Instead of climbing further up the marketing ladder, Elena found herself in an entirely new role: leading a cross-functional team of marketers, data scientists, and product developers in a position called "Consumer Experience Intelligence."

"It's not a promotion in the traditional sense," Elena explained. "It's a lateral expansion into a space that didn't exist before. My value isn't in managing more people or controlling bigger budgets—it's in my ability to integrate human creativity with machine intelligence across functions."

Elena embodies what Josh Bersin highlights as the rise of the Superworker—professionals who leverage AI to dramatically enhance their productivity, performance, and creativity, shifting from traditional roles to new positions that blend human expertise with machine intelligence (Bersin, 2025). Her journey underscores a strategic approach that involves repositioning at the intersection of human judgment and AI capabilities, advocating neither blind resistance nor uncritical acceptance, but rather a deliberate evaluation of the evolving landscape of value creation.

THE DIVISION OF LABOR: WHAT AI DOES VS. WHAT HUMANS MUST OWN

The first step to becoming an Augmented Professional is understanding the fundamental division of labor between human and machine intelligence. This isn't about competition—it's about complementary strengths.

What Ai Does Best Vs. What Humans Must Own

What Ai Does Best	What Humans Must Own
High-Volume Pattern Recognition Spotting regularities across massive datasets	**Novel Pattern Recognition** Identifying connections between seemingly unrelated domains
Prediction Based on Historical Data Forecasting outcomes using existing patterns	**Judgment in Ambiguity** Making decisions with incomplete information and unclear outcomes
Rapid Iteration Generating multiple variations quickly	**Value-Based Discernment** Determining which variations actually matter and why
Rules-Based Decision Making Applying consistent logic at scale	**Contextual Understanding** Interpreting situations considering cultural, ethical and human factors
Information Processing Analyzing structured data efficiently	**Meaning-Making** Creating narratives that build meaning from information
Answering Defined Questions Responding to clear queries	**Asking Essential Questions** Identifying which questions should be asked in the first place
Optimizing Known Metrics Maximizing defined success measures	**Defining Success Metrics** Determining what should be measured and why

The Augmented Professional doesn't compete with AI on the left column—that's a losing battle. They focus on mastering the right column, then strategically direct AI to handle the left.

The more of your work that lives in the left column, the more at risk you are.

This distinction isn't temporary. As AI capabilities advance, the left column expands to include more complex tasks, but the right column remains distinctly human. The value gap between those who understand this division and those who don't isn't narrowing—it's widening exponentially.

THE LIBERATION EFFECT: HOW AI EXPANDS HUMAN POTENTIAL

Lisa, a litigation attorney at a mid-sized law firm, used to spend roughly 30 hours per week on document review—identifying relevant evidence in mountains of emails, contracts, and corporate records. When her firm implemented an AI-powered document review system, her initial reaction was fear.

"I worried about my billable hours," she admitted. "Document review was 60% of my time. If that disappeared, what would happen to my position?"

Six months later, Lisa had the highest revenue generation in her department's history, despite logging fewer hours.

"The AI didn't make me irrelevant—it **liberated** me," she explained. "Instead of spending 30 hours staring at documents, I now spend 10 hours reviewing the AI's findings and 20 hours developing case strategy, preparing better arguments, and providing more meaningful counsel to clients."

In contrast, her colleague Mark resisted the AI tools, insisting on conducting document review "the proper way." As clients began questioning his invoices—"Why am I paying partner rates for work an AI could do?"—Mark's utilization rate dropped by 40%. He was eventually moved to a non-partner track position.

Lisa's experience demonstrates the true power of AI—not in what it replaces but in what it liberates:

- It liberates your **time** from low-value tasks
- It liberates your **attention** from routine problems
- It liberates your **creativity** from conventional constraints
- It liberates your **impact** from human-only limitations

This liberation effect creates a fundamental choice: Will you use this newfound freedom to merely work less, or will you use it to work at a higher level? The Augmented Professionals are choosing the latter, directing their liberated resources toward the work only humans can do.

FOUR COMMON ANTI-LADDER MISTAKES

As professionals navigate this transformed landscape, four common mistakes repeatedly undermine their transition to the new paradigm:

1. Waiting for Organizational Recognition Before Evolving

Many professionals see the AI shift happening but wait for their organization to formally redefine their role before adapting. They reason: "I'll change when my job description changes." This passive approach guarantees you'll be playing catch-up rather than leading transformation.

The Fix: Proactively seek AI-related responsibilities before they become mandatory. Volunteer for pilot programs, initiate AI-augmented projects, and demonstrate value in new ways without waiting for permission.

2. Defining Value by Tasks Completed, Not Problems Solved

When asked about their contribution, task-oriented professionals list what they've done: reports created, meetings facilitated, analyses completed. This task-orientation makes you highly vulnerable as AI increasingly handles execution.

The Fix: Reorient your professional identity around outcomes and problems solved rather than tasks completed. In conversations and documentation, emphasize the business impact of your work, not just the activities performed.

3. Building Depth Without Unique Combinations

Many professionals double down on specialized expertise without developing complementary skills that create unique value. While depth remains important, singular specialization creates vulnerability to AI disruption.

The Fix: Pair your deep expertise with complementary skills that create rare combinations. Financial expertise + ethical reasoning. Design talent + psychological insight. Technical knowledge + storytelling ability.

4. Learning AI Tools Without Reimagining Contribution

Some professionals diligently learn AI tools but apply them only to existing workflows, treating them as slightly better versions of previous technologies. This incremental approach misses the transformative potential of human-AI collaboration.

The Fix: As you learn AI tools, simultaneously reimagine your entire contribution. Don't just use AI to do your existing job faster—use it to redefine what your job could be.

THE CAREER ANTI-LADDER: SUCCEEDING IN A POST-HIERARCHICAL WORLD

The collapse of traditional career ladders isn't just about technology—it reflects a fundamental shift in how value is created in an AI-powered economy. When machines can handle routine execution at scale, the linear progression model based on gradually mastering standardized tasks becomes obsolete.

Instead of climbing predefined rungs, professionals now need to expand their capabilities in multiple dimensions simultaneously. This is what I call the "Career Anti-Ladder"—a progression model built for the AI era.

Instead of climbing upward, you expand outward through four distinct zones of increasing value:

Zone 1: The Efficiency Zone (Low-Value Territory)

This is where you use AI to do your existing job faster or better. It's a starting point, but it's low-value territory because everyone has access to the same tools. You're still defined by your job title and description—using AI merely as a productivity tool within your current boundaries.

At first glance, the shift from Efficiency to Extension may seem subtle—but it's the difference between working faster and working smarter. In Zone 1, AI helps you complete tasks more quickly within your existing role. In Zone 2, AI extends your skill set beyond your defined job, allowing you to step into higher-value work that wasn't previously part of your scope.

Warning: If you stay in Zone 1, you'll eventually be commoditized. You're competing with everyone else using the same AI tools for the same purpose.

Zone 2: The Extension Zone (Middle-Value Territory)

Here, you use AI to extend your capabilities beyond your defined role, actively blurring boundaries between functions. You're no longer defined by your job title but by your expanded impact across traditional departmental lines.

Example: A content marketer who uses AI to perform basic data analysis, allowing them to both create content and measure its effectiveness without involving the analytics team.

Zone 3: The Expertise Zone (High-Value Territory)

In this zone, you develop deep expertise in orchestrating AI systems to solve complex problems, becoming the go-to specialist who understands exactly how to direct AI for maximum effectiveness. You're defined by your ability to get results through human-AI collaboration.

Example: A financial analyst who builds custom AI workflows that combine multiple models to deliver insights no single system could provide, becoming the organization's specialist in AI-powered forecasting.

Zone 4: The Evolution Zone (Highest-Value Territory)

At this highest level, you're using AI to create entirely new types of value that weren't possible before. You're not just doing existing jobs better; you're inventing new forms of work. You're defined by your ability to envision and create what doesn't yet exist.

Example: An educator who designs a completely new learning model that combines AI-personalized curriculum with human mentorship, creating educational outcomes that neither could achieve alone.

THE CAREER ANTI-LADDER IN ACTION

The journey from Zone 1 to Zone 4 isn't a steady climb—it's a strategic expansion. It requires constantly pushing the boundaries of what's possible with human-AI collaboration.

Take Maya, a graphic designer who initially worried AI image generators would make her obsolete (Zone 1). Instead of competing with AI on technical execution, she developed expertise in directing these tools to translate complex brand guidelines into consistent visual outputs (Zone 3). Eventually, she created a new role for herself as an "AI-Human Creative Director," designing workflows that combined AI generation with human curation for large-scale visual projects that would have been impossible before (Zone 4).

In contrast, her former colleague Jason saw AI as an existential threat. "These tools will destroy the design profession," he insisted. Rather than learning to direct the AI tools, he focused on refining his manual design skills, believing clients would always prefer "authentic human creativity." As design automation accelerated, Jason found himself unable to compete on either price or scale. He's now teaching traditional design at a local community college, still avoiding AI tools.

Maya isn't climbing a career ladder—she's expanding her professional territory into spaces AI can't reach alone. Her value comes not from seniority but from her ability to create new forms of impact through human-AI collaboration.

STRATEGIC QUESTIONS FOR CAREER REINVENTION

To navigate this transformed landscape, professionals need a structured framework to evaluate their current position and plan their evolution. These strategic questions will help you identify your AI-era career path:

1. Where is Your Unique Human Value?

- In your current role, which aspects require distinctly human judgment, creativity, or ethical reasoning?
- When colleagues or clients seek your input, is it primarily for information (which AI could provide) or for judgment (which remains uniquely human)?
- What problems do you solve that require balancing competing values, considering subtle contextual factors, or making decisions with incomplete information?

2. What Would Remain If AI Automated 80% of Your Tasks?

- If an AI system could handle 80% of your current activities, which specific tasks would likely remain human-domain?

- Would these remaining responsibilities constitute a coherent role, or would you need to evolve your contribution?
- How might you expand these distinctly human elements to create sufficient value, even as AI handles more routine aspects of your work?

3. Where Is Your Industry's Human-AI Frontier?

- What new roles are emerging in your field that combine human judgment with AI capabilities?
- Which organizations in your industry are pioneering new approaches to human-AI collaboration?
- What capabilities are these frontier roles requiring that traditional positions don't emphasize?

4. What Unique Skill Combinations Could You Develop?

- What complementary capabilities could you pair with your existing expertise to create distinctive value?
- Which of your interests or secondary skills might combine with your primary expertise to create a unique perspective?
- What domains adjacent to your field are being transformed by AI in ways that create new connection opportunities?

The most successful professionals won't be those with the clearest advancement path but those who continuously reinvent their contribution as technology and organizational needs evolve.

THE 5 SKILLS THAT WILL NEVER BE AUTOMATED: YOUR AI-PROOF FOUNDATION

As machines become more capable, certain human skills become **more valuable**, not less. These aren't just any soft skills—they are specific human capabilities that become exponentially more important as AI handles increasingly complex technical tasks:

1. **Contextual Creativity**: Creating entirely new patterns rather than variations within existing frameworks

2. **Second-Order Emotional Intelligence**: Understanding and navigating complex human dynamics beyond simple emotion recognition

3. **Strategic Sensemaking**: Connecting seemingly unrelated developments into coherent strategic narratives

4. **Ethical Judgment**: Balancing competing values and making decisions with moral dimensions

5. **Adaptation Velocity**: Rapidly acquiring and applying new capabilities in changing circumstances

These skills form the foundation of your AI-proof professional identity. While specific technical abilities may be automated, these meta-capabilities create value precisely because they can't be reduced to algorithms.

The Augmented Advantage: Why the Future Belongs to AI Collaborators

The emerging evidence suggests that the highest performance comes not from humans alone or AI alone, but from their strategic combination.

This pattern appears consistent across domains: human-AI partnerships can achieve outcomes that neither could accomplish independently.

This "augmentation advantage" creates a widening gap between three groups of professionals:

1. **The Resisters** who avoid AI tools (rapidly losing ground)

2. **The Users** who employ AI as a productivity tool (maintaining relevance temporarily)

3. **The Augmented** who strategically integrate AI to transform how they work (gaining exponential advantage)

The gap between these groups isn't static—it's accelerating. While the resisters focus on protecting their traditional role and the users focus on incremental efficiency, the augmented professionals are reinventing entire approaches to value creation.

Action Box: Today's First Step

In the next five minutes, identify one AI-driven shift already happening in your field—and one action you can take this week to move beyond your current role. AI isn't waiting. Neither should you. Then, write down one thing you will do this week to move up the Anti-Ladder. Small shifts create momentum. Make yours today.

If you're still in Zones 1 or 2, identify one specific action you can take this week to move toward a higher zone. This single shift is your next step toward becoming an Augmented Professional.

As AI reshapes career trajectories, the question isn't just how to climb the ladder—it's whether a ladder even exists anymore. If traditional career progression is eroding, what replaces it? The answer lies not in credentials but in proof. In a world where AI can execute tasks faster than humans, the real differentiator is not what you claim to know—it's what you can demonstrate.

Now that you understand how AI is redefining career progression, let's explore why traditional notions of credentials and education have become increasingly misaligned with what creates real value in the AI era.

THREE THINGS FOR THIS WEEK

Reality Check: There's No Ladder—Only Leverage

1. Write down three skills you currently rely on. If AI could replicate two of them, what would be left?
2. Identify one area where you can **pivot or expand** your skill set to stay ahead of automation.
3. Research one real-world example of how AI has disrupted a career path similar to yours.

03

DEGREES ARE DEAD
PROOF OF WORK IS THE NEW CURRENCY

*AI doesn't care about credentials
—just what you can produce.*

James had followed the rules perfectly. Bachelor's degree from a respected university. Entry-level position at a Fortune 500 company. MBA from a top-20 program. Middle management by 35. All the checkboxes dutifully ticked on his march toward executive leadership.

Then a strategic restructuring eliminated his entire division. Not because the work wasn't needed, but because a combination of AI systems and a network of specialized contractors could deliver better results at half the cost.

"I built my entire career around climbing a ladder that no longer exists," James told me, still processing his shock six months later. "I was so focused on the next rung that I never noticed the whole structure was being dismantled."

But the collapse of traditional career paths isn't just about roles disappearing—it's about the old system of proving your worth becoming irrelevant. AI isn't just burning the rungs; it's making the entire credentialing system obsolete, forcing a shift toward proof-of-work over formal education.

James isn't alone. Across industries, the traditional career progression—that predictable path from entry-level to middle management to executive leadership—is collapsing. The clearest sign isn't layoffs or reduced hiring. It's the **disappearance of middle management itself**.

Many major companies are systematically flattening their organizational hierarchies, driven by advancements in AI and automation. These organizations are reducing management layers to streamline decision-making and improve efficiency, with significant restructuring efforts highlighting this shift as they downsize managerial roles in favor of AI-driven coordination and workflow automation.

The career ladder isn't just changing. It's fundamentally **dead. AI didn't burn down the career ladder—it automated the rungs one by one. The only way forward is to stop climbing and start stacking new skills.** The shift isn't theoretical—it's already playing out. Once-stable industries are experiencing a silent restructuring. The rise of AI, automation, and decentralized workforces has forced companies to rethink not just how work gets done, but who does it—and whether traditional roles even need to exist anymore. And while Chapter 2 explored how AI is burning the rungs of traditional career progression, there's an equally seismic shift happening in parallel: **the collapse of credentials as career currency.**

Traditional career paths—put in your time, master each role, wait your turn—have given way to more dynamic, nonlinear trajectories. But something even more fundamental is changing: the very basis upon which professional opportunity is distributed. The rules that governed career advancement for decades have been rewritten. Those still playing by the old rulebook aren't just at a disadvantage—they're participating in a game that no longer exists.

This chapter isn't about mourning the death of credentials. It's about embracing the far more exciting reality that's replaced it: an era where what you can demonstrably create matters infinitely more than the degrees on your wall or the lines on your resume.

THE EDUCATION APOCALYPSE: WHY CREDENTIALING IS COLLAPSING

A college degree—once the golden ticket to middle-class stability—has become simultaneously more expensive and less valuable. Consider these alarming realities:

- **Escalating Student Loan Debt**: The average federal student loan debt per borrower in the United States reached $38,375 by late 2024, with total balances (including private loans) hitting $41,520, reflecting a 66% rise over the past decade (Education Data Initiative, 2025).
- **High Underemployment Rates**: About 52% of bachelor's degree holders aged 25-34 are underemployed, working in jobs not requiring their degree, signaling a persistent

mismatch between education and employment (Strada Education Network, 2023).

- **Future Job Market Disruption**: The World Economic Forum's *Future of Jobs Report 2023* predicts that 23% of global jobs will face significant disruption by 2027, with 44% transformed by technologies like AI and 14% lost entirely, emphasizing the urgent need for education systems to adapt to a workforce where skills evolve faster than traditional degrees can accommodate (World Economic Forum, 2023).

These aren't just concerning trends—they're symptoms of a fundamental disconnect between traditional higher education and the AI-powered workforce of the future.

The educational system is still operating as if the economy moves at the same pace it did decades ago. But while degrees take four years to complete, AI upends industries in months. This disconnect is leaving graduates unprepared and employers looking elsewhere for real skills.

The hard truth? **The entire credential-based education model is collapsing in real time.**

This isn't just about whether college is "worth it" financially. It's about a deeper structural failure: Formal education's inability to keep pace with the accelerating rate of skill evolution in the marketplace.

When the half-life of professional skills has shrunk from decades to months, a four-year degree program becomes obsolete before it even concludes. By the time students graduate, they've already fallen behind those who've been learning through direct market engagement.

Traditional education isn't just overpriced—it's fundamentally misaligned with how value is now created and measured.

This educational misalignment isn't just affecting individual career outcomes—it's undermining the very institutions that have traditionally served as career gatekeepers. As the disconnect between credentials and capability widens, even elite universities are finding their century-old business models increasingly vulnerable.

But What About Medicine, Law, and Engineering?

Not every industry follows the same trajectory. Fields like **medicine, law, and engineering** still require **formal credentials, licensure, and regulated pathways** for good reason—public safety, ethical considerations, and professional accountability demand it. You wouldn't want a self-taught surgeon, and for now, AI isn't replacing human expertise in **courtrooms, hospitals, or structural engineering**.

However, even in these fields, the landscape is shifting. AI is already **augmenting** rather than replacing professionals—AI-assisted diagnostics in medicine, legal research automation, and AI-powered engineering simulations. While degrees still serve as **gateways** in these professions, the professionals thriving within them are the ones **adopting AI, developing hybrid skill sets**, and **proving their ability to innovate** within their fields.

The key takeaway? **If you're in a regulated profession, a degree is still essential—but in most industries, proof of work has overtaken credentials as the new standard of value.**

THE COMING COLLAPSE OF UNIVERSITY PRESTIGE

For centuries, elite universities have functioned as gatekeepers of opportunity. Their brands have signaled quality, conferring advantages regardless of the actual skills graduates acquired. But in an AI-powered economy where proof of work trumps pedigree, this system is unraveling.

Three forces are accelerating this collapse:

1. AI Exposes Outdated Curricula

Universities are inherently slow-moving institutions—curricula take years to develop and receive approval. Meanwhile, AI and automation are transforming industries at an exponential rate. This widening gap between what universities teach and what the market demands undermines the core value proposition of these institutions.

A four-year computer science curriculum developed in 2021 will be teaching programming paradigms in 2025 that were already outdated when the curriculum was approved. Meanwhile, self-taught engineers are mastering emerging technologies like AI-generated software development and blockchain applications in real-time, often outpacing degree-holders in market relevance. How can a static educational program compete with AI-powered learning systems that adapt in real-time to technological and market changes?

2. The Employer Shift

Employers are steadily reducing degree requirements in job postings, pivoting to skills-based hiring to address talent shortages and evolving job demands.

A February 2024 analysis by Indeed Hiring Lab found that the proportion of U.S. job postings requiring a bachelor's degree fell from 20.4% in January 2019 to 17.8% by January 2024, with 52% of postings in 2024 listing no education requirement at all (Indeed Hiring Lab, 2024).

Major tech firms are driving this change, with Google reporting that its Career Certificates, launched in 2020 and expanded through 2023, have helped over 300,000 U.S. learners secure jobs without degrees by mid-2023, often in roles like IT support (Grow with Google, 2023).

Similarly, IBM's 2023 Impact Report notes that 20% of its U.S. hires in 2023 lacked four-year degrees, a rise from 15% in prior years, reflecting its "New Collar" focus on skills like coding over formal education (IBM, 2024).

In specialized fields like AI and sustainability, research shows a 30% increase in job postings prioritizing certifications or project experience over degrees between 2020 and 2023, based on an analysis of over 10,000 listings (Bone et al., 2024).

This shift signals a broader move toward valuing demonstrated ability, challenging the traditional reliance on academic credentials.

3. The Tuition Bubble

Rising tuition costs have already outpaced earning potential in many fields, creating an unsustainable financial model. As more professionals succeed without degrees—or with degrees from less prestigious, more affordable institutions—the return on investment for elite education will continue to deteriorate.

This creates a death spiral: decreasing ROI leads to fewer qualified applicants willing to pay premium prices, which forces institutions to either lower standards (diluting the brand) or raise prices further (exacerbating the ROI problem).

Within the next decade, even elite universities will face a stark choice: transform into skill incubators that provide modular, just-in-time education aligned with market needs, or risk irrelevance as their prestige premium evaporates. The institutions that survive won't be selling degrees—they'll be selling demonstrable capabilities and networks.

The institutions most threatened by this shift aren't responding with the radical reinvention required. Instead, they're doubling down on credentials—offering more specialized certificates, micro-degrees, and badges that attempt to signal relevance while preserving the fundamental credential-based model.

This approach misses the central reality: In an AI-powered economy, proof of work has definitively replaced proof of credential as the primary currency of professional opportunity.

THE RISE OF THE PROOF-OF-WORK ECONOMY

Elena never finished her computer science degree. Three semesters in, she landed an internship at a fast-growing startup where she built her first machine-learning model. The company offered her a full-time role, and she never returned to complete her degree.

Five years later, Elena leads an AI implementation team at a major technology company. Her compensation exceeds that of most of her former classmates who dutifully completed their degrees. More importantly, she's accumulated practical experience that no classroom could have provided.

"The degree wasn't delivering what I needed fast enough," she explained. "While my classmates were studying theoretical concepts that would be outdated by graduation, I was solving real problems with cutting-edge tools. My portfolio of completed projects has opened far more doors than a degree ever could have."

Elena's experience illustrates the fundamental shift underway: We're witnessing the rise of a proof-of-work economy, where demonstrated capability trumps formal credentials.

This isn't just a minor adjustment to how we evaluate talent. It's a fundamental inversion of the mechanisms that have governed access to professional opportunity for generations.

We're transitioning from a world where credentials opened doors to one where tangible impact keeps them open. Artificial intelligence models like OpenAI's GPT-4 prove their worth through output, not degrees, and employers are increasingly applying this lens to people. Here's how this plays out across sectors:

- **Technology:** IBM and Google have cut degree requirements for many roles, with IBM reporting 20% of its 2023 U.S. hires lacked four-year degrees and Google's Career Certificates placing over 300,000 learners in jobs by mid-2023, emphasizing skills over pedigrees (IBM, 2024; Grow with Google, 2023).

- **Finance:** Wall Street firms are testing analytical skills directly, with a 2024 survey showing that 35% of financial employers now value practical assessments over academic credentials, up from 28% in 2022 (Burning Glass Institute & Strada Education Network, 2024).
- **Marketing:** Agencies increasingly favor candidates with proven content creation, as 40% of U.S. creative job postings in 2023 sought portfolios or experience over degrees, a rise from 32% in 2021 (LinkedIn, 2024).

This shift reflects a growing consensus that traditional education often fails to signal the skills needed in today's fast-changing economy, pushing employers to ask: *What can you do, and how fast can you adapt?*

THE THREE TIERS OF PROOF: YOUR ADVANCEMENT ROADMAP

Not all proof of work carries equal weight. In the emerging opportunity marketplace, evidence of capability exists in a clear hierarchy—and understanding how to advance through these tiers is essential for career success.

Tier 3: The Passive Portfolio (Basic)

What It Is: This lowest tier consists of work samples, process documentation, and testimonials about past accomplishments. While better than credentials alone, this passive evidence has limited impact because it's backward-looking and often curated to show only successes.

Example: A UX designer with a portfolio website displaying final designs but lacking context about problems solved, constraints navigated, or metrics improved.

Next Steps to Level Up:

- Document your thinking process, not just the final products
- Share both successes and instructive failures
- Create content explaining your approach and methodology
- Engage with feedback rather than just displaying work

Tier 2: The Public Challenge (Intermediate)

What It Is: This middle tier involves publicly solving problems, participating in competitions, or creating open work that others can evaluate and build upon. This carries more weight because it involves public accountability and real-time performance rather than carefully selected samples.

Example: A data scientist participating in Kaggle competitions, publishing approaches, collaborating with others, and documenting both successes and failures in the public arena.

Next Steps to Level Up:

- Move beyond responding to pre-existing challenges
- Identify unmet needs others haven't recognized
- Build a public following for your approach
- Create educational content that helps others solve similar problems

Tier 1: The Created Opportunity (Advanced)

What It Is: The highest tier involves identifying problems others haven't recognized and creating solutions that demonstrate both capability and initiative. Rather than solving assigned problems, this involves defining new value categories and demonstrating the ability to deliver against them.

Example: A marketing professional who identifies an unaddressed need for connecting sustainability metrics to brand engagement, develops a methodology, creates implementation tools, and publishes case studies demonstrating effectiveness.

Next Steps to Scale Your Impact:

- Convert your solution into frameworks others can adopt
- Develop educational pathways to help others implement your approaches
- Build communities around the problems and solutions you've identified
- Create systems allowing your impact to scale beyond your direct effort

The professionals commanding premium opportunities aren't just showcasing their work—they're creating evidence at the highest tier, demonstrating not just technical execution but strategic vision and self-directed initiative.

Consider Maya, a designer who wanted to transition into AI-assisted creativity. Rather than simply taking courses or building a portfolio, she identified a gap in how fashion brands were utilizing AI, created a speculative project demonstrating her approach, published her process as open research, presented at industry events, and built tools that other designers could use. Within six months, she had multiple job offers from fashion technology companies, despite having no formal credentials in AI or fashion design.

Moving up these tiers isn't just about creating more impressive work—it's about fundamentally changing your relationship with the market. As you advance from passive showcasing to active opportunity creation, you position yourself as a value creator rather than just a skill possessor.

THE PROOF-OF-WORK PORTFOLIO: YOUR NEW CREDENTIAL

If degrees are dead, what replaces them? The answer is the proof-of-work portfolio—a dynamic, evolving body of evidence that demonstrates your capabilities in real-time.

Unlike traditional resumes or credentials, which are static and backward-looking, the proof-of-work portfolio is constantly evolving, capturing both completed work and ongoing development. It serves as a living testament to what you can do and how you approach challenges. These portfolios provide a **more accurate and comprehensive** representation of a candidate's skills and competencies, offering tangible evidence of their ability to perform in real-world scenarios (Forbes, 2024).

THE FOUR ESSENTIAL COMPONENTS

An effective proof-of-work portfolio contains four key elements:

1. Tangible Artifacts

These are the actual outputs of your work—code repositories, design samples, writing samples, project documentation, data analyses, etc. The key is that these must be accessible and reviewable, not just described.

2. Process Documentation

This reveals how you think and work, not just what you produce. It includes documentation of your approach, the challenges you faced, the decisions you made, and what you learned.

3. Impact Evidence

This connects your work to measurable outcomes and real-world value. It answers the question: "So what?" about your capabilities.

4. Learning Trajectory

This demonstrates your capacity for growth and adaptation— perhaps the most valuable capability in a rapidly changing environment.

The portfolio approach represents a fundamental shift in how we think about professional identity. Rather than being defined by credentials or job titles, you become defined by the tangible evidence of what you can create and how you create it. This shift gives you more control over your professional narrative and creates more direct connections between your capabilities and opportunities.

BUILDING YOUR PROOF-OF-WORK PORTFOLIO

The most effective portfolios aren't created after the fact—they're built as a deliberate by-product of your work. Here's how to begin developing yours:

1. **Documentation by default**: Make documenting your process and outcomes a standard part of every project, not an afterthought.
2. **Public by default**: Unless confidentiality is required, make your work publicly accessible. The visibility itself adds credibility.
3. **Narrative connection**: Don't just showcase isolated projects—create explicit connections between them that reveal your development and specialization.
4. **Continuous curation**: Regularly review and update your portfolio, ensuring it reflects your current capabilities and interests, not just historical work.
5. **Stakeholder orientation**: Organize your portfolio around the problems you solve and the value you create, not just the skills you possess.

Consider Rajiv, an operations professional who transformed his career by systematically documenting his process improvements at a manufacturing company. While his colleagues saw documentation as a burdensome administrative task, Rajiv approached it as portfolio-building—capturing not just what changed but why it mattered and how he approached the problem. When an economic downturn led to layoffs, Rajiv's evidence-rich portfolio landed him a consulting role at twice his previous salary. His proof-of-work spoke volumes that his resume alone never could.

THE SELF-EDUCATED PROFESSIONAL: LEARNING WITHOUT INSTITUTIONS

If traditional educational institutions are increasingly misaligned with market needs, how should professionals approach continuous learning? The answer lies in what I call "just-in-time education"—a self-directed approach to learning that prioritizes immediate application over comprehensive curricula.

Andrew, a marketing professional, needed to develop data science capabilities to remain relevant in his increasingly analytical field. Rather than enrolling in a data science degree program, he:

1. Identified a specific analytical challenge in his current role

2. Learned exactly what he needed to solve that problem

3. Applied his learning immediately to create measurable impact

4. Built on that foundation for the next challenge

5. Repeated this cycle, building capabilities through direct application

"I essentially created my own applied data science education," Andrew explained. "But instead of following a predefined curriculum, I let actual challenges drive my learning. The result was faster progress and immediate value creation instead of deferred application."

This approach represents a fundamental shift in how learning relates to work:

Traditional Education	Just-In-Time Education
Comprehensive first, application later	Application first, learn as needed
Institution-directed curriculum	Problem-directed curriculum
Periodic, intensive learning periods	Continuous, integrated learning
Learning separated from work	Learning embedded in work

CREDENTIALS AS THE PRIMARY OUTPUT SOLUTIONS AS THE PRIMARY OUTPUT

This shift doesn't mean formal education has no value. Rather, it suggests a new relationship between institutional learning and professional development—one where formal education serves as a foundation that's continuously extended through self-directed, application-focused learning.

The most successful professionals are becoming increasingly sophisticated in how they navigate learning resources, combining elements of:

- **Focused courses** for specific skill development
- **Communities of practice** for tacit knowledge and feedback
- **Applied projects** for immediate integration and testing
- **Mentorship relationships** for contextual guidance
- **Self-directed research** for customized knowledge development

The key insight is that learning itself is becoming a meta-skill—the ability to rapidly acquire new capabilities as needed is now more valuable than possessing any specific set of skills or credentials.

CASE STUDIES IN CREDENTIAL-FREE SUCCESS

The Self-Taught AI Specialist

Michael had spent seven years as a marketing analyst at a consumer packaged goods company when generative AI began transforming his industry. Rather than pursuing a formal AI education, he took a different approach:

1. He identified a specific application of generative AI that could transform his company's consumer research
2. He learned just enough about prompt engineering and LLM capabilities to create a prototype
3. He tested his solution with real data and demonstrated measurable improvements
4. He documented his process and results, creating tangible proof of his new capabilities

Within three months, Michael had established himself as his department's AI specialist, despite having no formal credentials in the field. Six months later, he launched an internal AI strategy team that now commands a seven-figure budget.

"I never set out to become an AI expert," Michael explained. "I just solved a specific problem using AI tools, then another, then another. The expertise emerged through application, not through certification."

Michael's experience highlights a critical reality: In emerging fields, practical application often outpaces formal education. Those waiting for credentials before applying new technologies find themselves perpetually behind those who learn through direct engagement.

The Open Source Authority

James was a self-taught developer who never completed college. Recognizing that his lack of credentials might limit his opportunities, he:

1. Contributed consistently to open-source projects, creating a public record of his capabilities
2. Documented his learning process and technical decisions on a personal blog
3. Created educational content helping others solve problems he had overcome
4. Built a specialized tool that addressed a common pain point in his field
5. Became an active participant in developer communities, offering assistance and insight

Today, James leads engineering at a technology startup, earning significantly more than many credentialed peers. His GitHub profile and contribution history have served as a more powerful credential than any degree could provide.

"Open source created a meritocracy where my work could speak for itself," James explained. "Anyone can verify not just what I've built, but how I think, how I collaborate, and how I've developed over time. That transparency is worth more than any credential."

James's path highlights how public work in transparent environments can establish authority and opportunity access without traditional credentials.

These cases aren't outliers—they represent an accelerating trend across industries. The individuals who thrive aren't waiting for institutional validation. They're creating proof that makes such validation unnecessary, building their own paths to opportunity through demonstrated capability rather than certified potential.

The Credential Bubble: Why Formal Education Must Transform or Die

The growing gap between traditional education and market needs signals a "credential bubble"—a system where degrees are overvalued relative to their practical utility. Consider the evidence:

- Tuition at public 4-year institutions has risen 136% above inflation from 1971-72 to 2021-22, outpacing economic fundamentals (National Center for Education Statistics, 2023).
- Total U.S. student loan debt reached $1.727 trillion by late 2023, fueling degree pursuits despite uncertain returns (Federal Reserve Bank of New York, 2024).
- Undergraduate enrollment dropped 8% from 15.975 million in 2019 to 14.697 million in 2022, reflecting skepticism about degree value beyond a pandemic blip (National Student Clearinghouse Research Center, 2023).

This isn't a sudden collapse but a slow reckoning: 52% of recent graduates work jobs not requiring degrees, per 2023 data, pushing employers toward skills over credentials (Strada Education Network, 2023). Formal education must adapt—or risk irrelevance.

For the credential bubble to deflate without catastrophic consequences, educational institutions must radically transform their models to align with the realities of the proof-of-work economy. This means:

- **Embedding real-world application throughout the learning process**, not just at its conclusion
- **Creating visible evidence of capabilities** as a core output, not just an informal by-product
- **Establishing direct market connections** that enable immediate value creation
- **Shifting from comprehensive to modular education** that can be assembled according to individual needs
- **Embracing continuous assessment** based on demonstrated capabilities rather than time-based progression

Some forward-thinking institutions are already moving in these directions. Organizations like Western Governors University have pioneered competency-based education that focuses on demonstrated mastery rather than credit hours. Coding bootcamps like Lambda School (now Bloom Institute of Technology) have introduced income share agreements that align institutional success with graduate outcomes.

But these innovations remain exceptions rather than the rule. Most traditional institutions continue operating as if the fundamental credential model remains sound, making only incremental adjustments to an increasingly misaligned system.

The institutions that will thrive aren't those with the strongest historical brands but those that most effectively bridge the gap between learning and application—creating not just knowledgeable graduates but professionals with compelling evidence of their capabilities.

The New Rules of Professional Advancement

The shift from credential-based to proof-based professional advancement requires new strategies for building and navigating careers. Here are the five essential principles for thriving in this new landscape:

1. Create Evidence, Don't Just Acquire Skills

In the credential economy, skill development was enough—the degree or certificate served as proxy evidence. In the proof-of-work economy, the evidence itself is the product.

Strategy: For every significant capability you develop, create tangible, visible evidence of that capability that exists independently of any employer or institution.

2. Learn Through Application, Not Just Absorption

Traditional education separates learning from application—first you learn, then you apply. In the proof-of-work economy, learning through direct application accelerates both capability development and evidence creation.

Strategy: Structure learning around actual problems rather than abstract curricula, ensuring immediate application of new knowledge.

3. Build in Public, Not Just in Private

The credential economy privatized learning and development— your growth was between you and the institution. The proof-of-work economy rewards public learning that creates visible evidence of both capabilities and meta-learning skills.

Strategy: Document your learning and development process publicly, sharing not just successes but challenges, failures, and growth.

4. Solve Real Problems, Not Just Theoretical Exercises

Educational institutions rely heavily on artificial exercises and simulations. The proof-of-work economy values solving actual problems that create real value, even at small scales.

Strategy: Seek out genuine problems to solve rather than theoretical exercises, creating evidence connected to actual outcomes.

5. Own Your Verification, Don't Outsource It

The credential economy outsourced verification to institutions— your capability was assumed based on their certification. The proof-of-work economy requires professionals to make their capabilities independently verifiable.

Strategy: Structure your evidence to be self-verifying or independently verifiable without relying on institutional authority.

These principles represent not just tactical adjustments but a fundamental shift in how professionals develop and demonstrate their value. The most successful participants in the proof-of-work economy internalize these principles as core operating assumptions rather than occasional strategies.

THE 30-DAY PROOF-OF-WORK CHALLENGE

Understanding these principles isn't enough—you need to take immediate action to begin building your proof-of-work portfolio. The following 30-day challenge provides a structured approach to get started:

Week 1: Identify Your Proof-of-Work Focus

- Choose one high-value problem in your industry that you can start working on
- Research existing solutions and find an angle that others are missing
- Map your existing capabilities against the problem requirements
- Identify any learning gaps you'll need to address

Week 2: Build and Document Your Work

- Start your first public case study (a Medium post, Twitter thread, video breakdown, GitHub project, etc.)
- Document your process as you go—not just what you're doing, but why
- Show your thinking, including false starts and changes in direction
- Create tangible artifacts that demonstrate your capabilities

Week 3: Engage and Iterate

- Share your work-in-progress with professionals in your field
- Actively seek feedback from people whose opinion you respect

- Refine and adapt based on real-world input
- Demonstrate your learning agility by implementing improvements

Week 4: Publish and Amplify

- Package your work into a compelling portfolio piece
- Distribute across platforms (LinkedIn, Twitter, relevant forums, niche communities)
- Engage in discussions and start connecting with key people
- Begin planning your next proof-of-work project based on what you've learned

Key Milestones:

- Day 7: Problem selected and research completed
- Day 14: First draft of your work and documentation completed
- Day 21: Feedback collected and improvements implemented
- Day 30: Final work published and amplification strategy in action

By completing this 30-day challenge, you'll have taken the crucial first step from credential-based to proof-based professional advancement. More importantly, you'll have established the habits and systems needed for ongoing portfolio development.

Waiting for the world to validate your skills is the surest way to get left behind. AI isn't asking for permission—it's proving itself through results. You need to do the same.

The urgency of this challenge cannot be overstated. While most professionals continue operating under old assumptions, the marketplace is already rewarding those with visible, verifiable proof of their capabilities. The window to get ahead of this shift is closing—those who wait for institutional permission will find themselves increasingly disadvantaged.

NAVIGATING THE TRANSITION: FROM CREDENTIALS TO PROOF

The shift from credentials to proof isn't happening overnight, and we're currently in a transition period where both systems coexist. This creates both challenges and opportunities for professionals at different career stages.

For Students and Early Career Professionals

You face the most difficult choice: invest in traditional credentials that may depreciate before you complete them or focus on building proof-of-work that may not be universally recognized yet.

Strategic approach:

1. If pursuing formal education, choose programs that produce tangible evidence alongside credentials
2. Develop a parallel proof-of-work portfolio that demonstrates practical capabilities
3. Seek opportunities to solve real problems during your education, not just after it
4. Build relationships with professionals who recognize and value proof over credentials
5. Be prepared to demonstrate your capabilities directly rather than relying on credentials to open doors

For Mid-Career Professionals

You likely have credentials but may lack sufficient proof-of-work outside your current employer. Your challenge is creating portable evidence without jeopardizing your current position.

Strategic approach:

1. Identify capabilities that are valuable but invisible in your current role
2. Create side projects that demonstrate these capabilities without competing with your employer
3. Document and share your problem-solving approach, not just outcomes
4. Build a narrative that connects your formal credentials with your practical experience
5. Participate in public professional communities where you can demonstrate expertise

For Senior Professionals and Executives

You may have relied heavily on reputation and credentials to advance. Your challenge is demonstrating adaptability and continued relevance in rapidly evolving domains.

Strategic approach:

1. Focus on creating evidence of your learning agility and adaptability, not just past accomplishments
2. Participate visibly in emerging conversations rather than only established ones
3. Create evidence of your ability to integrate new approaches with proven experience
4. Mentor others publicly, demonstrating both expertise and leadership

5. Develop and share frameworks that show your systematic approach to new challenges

ACTION BOX: YOUR FIRST STEP

Today: Begin your proof-of-work inventory

Take 5 minutes right now to identify one high-value capability you possess that lacks sufficient visible evidence. Then schedule 60 minutes this week to begin creating tangible proof of that capability—whether through a blog post, code repository, design sample, or case study.

The credential bubble won't deflate gradually—it will burst. When it does, will you be holding paper or proof?

The future won't ask for your résumé—it will demand your relevance. AI doesn't care about your degree, your job title, or how long you've been in the industry. It only cares about what you can create, solve, and contribute today. The professionals who cling to outdated credentials are **fighting gravity**—the ladder is already gone, and the only way forward is to build something new.

The question isn't whether proof-of-work will replace credentials—it already has. The only question is whether **you will be the one proving your value, or watching from the sidelines as others do.**

THREE THINGS FOR THIS WEEK

Reality Check: Your Resume Is a Receipt—Your Work Is the Product

1. Audit your resume or portfolio—does it prove your impact or just list credentials? Rewrite one section to showcase proof of work.
2. Build and share a **small proof-of-work project** in your field—even if it's just a LinkedIn post sharing insights.
3. Research one alternative learning path (certifications, bootcamps, apprenticeships) that could accelerate your growth faster than traditional education.

04

The Security Trap
Why Playing It Safe is the Biggest Risk

The illusion of safety has held people back long before AI. This chapter dismantles it.

When Karen lost her position as a senior marketing manager after 18 years at the same company, the first words out of her mouth were, "But I thought I was secure."

The CEO had personally assured her that her job was safe just three months earlier. Her performance reviews had been consistently stellar. She had built what she believed was an irreplaceable network of relationships both inside and outside the organization.

None of it mattered when an AI-powered marketing platform eliminated 60% of her department's workload overnight.

"I had built my entire life around the illusion of stability," Karen told me. "I chose the safe path at every junction. I stayed at one company rather than risk job-hopping. I specialized in one domain rather than exploring diverse paths. I prioritized institutional loyalty over market value."

Karen's story isn't unique. What makes it instructive is her reflection after six months of perspective: "The most dangerous decision I ever made was believing in job security. That belief didn't just cost me a job—it cost me nearly two decades of growth opportunities."

Karen eventually realized that her greatest mistake wasn't failing to predict disruption—it was failing to prepare for it. Six months later, she wasn't just job-hunting. She was rebuilding—pivoting into a role that combined her deep industry knowledge with newly acquired AI management skills, creating hybrid expertise that made her more valuable than before the layoff. "I'm not just employable again," she told me a year after losing her job. "I'm disruption-proof in a way I never was during those 18 'secure' years."

Karen's experience reflects a fundamental truth about human psychology—we're wired to seek stability and filter out signals of change, even when those signals are in plain sight. Our brains evolved to favor the familiar, creating a dangerous blind spot in times of rapid transformation.

The hard truth is this: Job security has always been a comforting fiction. A story we tell ourselves to feel safe in an inherently uncertain world. AI isn't creating workforce instability—it's merely exposing the instability that was always there.

And here's the contrarian truth that will transform your career: Uncertainty isn't your enemy. It's your greatest competitive advantage.

The problem isn't risk itself—it's that most people calculate risk using the wrong equation. They assume safety means keeping what they have when in reality, safety depends on staying valuable as the world shifts. AI isn't the real disruptor here—stagnation is. That's why we need a new way to measure risk. Instead of asking, What do I stand to lose if I make a move?, the real question is: What do I stand to lose if I don't?

THE RISK VS. REGRET EQUATION: A NEW WAY TO THINK ABOUT CAREER DECISIONS

Most professionals evaluate career choices through the lens of risk—what could go wrong if they take an uncertain path? But this is a fundamentally flawed calculation. The more important metric isn't risk—it's regret: What opportunities are permanently lost by staying put?

Introducing the Risk vs. Regret Equation

Career Value = (Potential Upside × Probability of Success) − (Immediate Risk × Severity) + Regret Prevention Value

This equation isn't just an abstract concept—it's a powerful mental model that reshapes how you evaluate career choices. It forces you to quantify not just the downside of risk, but the cost of inaction.

For most people, "immediate risk" dominates decision-making. They overweight short-term stability and underweight potential upside and regret prevention. This leads to systematically poor choices, particularly in a world where industries shift, skills become obsolete, and opportunities are fleeting.

Applying the Risk vs. Regret Lens: Meet Jordan

To see this in action, meet Jordan.

At 32, Jordan has spent the last eight years working in a well-established corporate role. The job is stable, pays well, and offers good benefits. But it's also stagnant—there's little room for upward mobility, and the work has become increasingly routine.

Then, an opportunity arises: A fast-growing startup in his industry offers him a role. The startup is promising but risky. It's still in early funding rounds, and while the position offers a higher long-term upside, it comes with lower short-term security. If the startup succeeds, Jordan could gain equity, develop valuable skills, and accelerate his career. But if it fails, he could be job-hunting within a year.

Jordan now faces a classic career dilemma. Should he play it safe and stay in his current role? Or should he take a leap into the startup world? To help him decide, he applies the Risk vs. Regret Equation:

Decision	Traditional Risk Focus	Risk vs. Regret Focus
Staying at a stable but stagnant job vs. joining a growing startup	Emphasizes the risk of startup failure (immediate job loss).	Weighs risk against regret of missing skill development, industry shifts, and equity upside.

Decision	Traditional Risk Focus	Risk vs. Regret Focus
Specializing deeply in one domain vs. developing cross-disciplinary expertise	Focuses on risk of being a "jack of all trades, master of none."	Considers regret of having non-transferable skills if the primary domain becomes obsolete.
Prioritizing immediate salary vs. learning opportunity	Emphasizes the risk of lower current income.	Calculates long-term income potential and future regret of stunted career growth.

Using actual numbers, Jordan applies the equation to his decision:

Scenario 1: Staying in His Current Role (Stable Job)

- Potential Upside: $50,000 (incremental career growth over 5 years)
- Probability of Success: 90% (his job is very secure)
- Immediate Risk: $10,000 (possible missed promotions or salary stagnation)
- Severity: 2 (low risk, since he maintains stability)
- Regret Prevention Value: $5,000 (moderate, since he knows his skills may slowly become outdated)

(50,000 × 0.9) - (10,000 × 2) + 5,000 = 30,000

Scenario 2: Joining the Startup

- Potential Upside: $200,000 (higher salary growth and stock options over 5 years)
- Probability of Success: 50% (startups are unpredictable)

- Immediate Risk: $30,000 (lower salary, risk of job loss)
- Severity: 7 (higher risk—startups fail frequently)
- Regret Prevention Value: $50,000 (very high—if the startup succeeds, he gains new skills, equity, and industry credibility)

$(200{,}000 \times 0.5) - (30{,}000 \times 7) + 50{,}000 = -60{,}000$

At first glance, this suggests that the safe choice (staying) is better. But let's reconsider.

- Personal Growth & Learning: While the startup has high risk, it also offers the potential to build in-demand skills, which could boost future career prospects far beyond his current role.
- Industry Trends: If his current industry is shrinking or automating, staying could be riskier than it appears.
- Opportunity Cost of Staying: The real cost of staying isn't just the lower salary growth—it's the stagnation in learning and network expansion.

Jordan realizes that if he stays, he may feel safer in the short term, but in five years, he could deeply regret not taking a chance. Armed with this new perspective, he chooses to join the startup, knowing that even if it fails, he will have gained valuable skills and experiences that make him more employable in the future.

WHY STABILITY IS AN ILLUSION

The pull of stability is strong because it feels rational. The world rewards you for following the script—until the script itself becomes obsolete.

But let's be clear: It's not a lack of intelligence that keeps people stuck.

It's fear—disguised as logic.

The problem? Logic built on a failing system is just a slower way to lose.

In a world where technological shifts happen overnight and entire industries are disrupted in months, the real risk isn't trying something new—it's betting your future on a script written for a world that no longer exists.

The Risk vs. Regret Equation forces you to ask:

What is the real cost of staying where I am?

Because the real question isn't, "What if I fail?"

It's, "What if I never try?"

THE SECURITY PARADOX: WHY "SAFE" CHOICES ARE NOW THE RISKIEST

Everything you've been taught about career "safety" is dangerously outdated.

Meet Daniel.

When he was offered a position at a prestigious law firm, it felt like the obvious secure choice—high salary, clear partnership track, and the prestige of working in an established institution. At the same time, he had another offer: a legal technology startup that was tackling AI-driven contract analysis. It seemed interesting, but also uncertain.

Daniel played it safe. He chose the traditional law firm, prioritizing stability over experimentation.

Two years later, the illusion of security shattered.

His firm eliminated 30% of its associate positions after AI systems dramatically reduced the need for document review and legal research. Meanwhile, the startup he passed on had skyrocketed, creating new high-impact roles for professionals who understood both legal practice and technology.

When Daniel finally reconnected with that legal tech startup two years later, it wasn't as a job applicant—it was as an outsider looking in, realizing he had missed his greatest professional opportunity. "I thought I was making the safe bet," he admitted. "Turns out, I was just betting on a world that no longer exists."

Daniel's experience highlights the Security Paradox—the increasingly common situation where choices made in pursuit of stability actually increase vulnerability to disruption.

For much of the 20th century, professional life was structured around a simple contract: loyalty in exchange for security. Those who stayed in their roles, built expertise, and advanced within a single company were rewarded with long-term employment and financial stability.

That world no longer exists.

Decades of globalization, automation, and corporate restructuring have undermined career stability, and AI-driven shifts are accelerating the pace. The riskiest move isn't embracing change—it's clinging to outdated career models assuming past patterns hold.

Adaptability, not avoidance, builds security in a landscape where industries are rapidly redefined. Here's the evidence:

- Median job tenure for U.S. wage and salary workers fell from 4.7 years in 2000 to 4.1 years in 2022, reflecting shorter career spans (U.S. Bureau of Labor Statistics, 2022).
- The average lifespan of S&P 500 companies shrank from 33 years in 1964 to 24 years by 2016, with forecasts nearing 20 years as of 2021, driven by market churn (Innosight, 2021).
- Alternative work arrangements grew from 10.7% of U.S. workers in 2005 to 15.8% in 2015, signaling a shift to flexible, skill-based roles (Katz & Krueger, 2019).

These trends—declining tenure, shorter corporate lifespans, and rising gig work—show stability was a historical anomaly. Success now hinges on adaptability and proven skills, not fixed roles. Netflix thrived by pivoting to streaming in 2007, while Blockbuster's hesitation led to bankruptcy in 2010—a stark lesson in the cost of resisting disruption.

The security trap works like this: You sacrifice growth, autonomy, and market value in exchange for perceived stability. You make yourself smaller to fit within the boundaries of a specific role. You identify with your position rather than your capabilities. You become dependent on a single source of income and validation.

Then, when the inevitable disruption occurs—whether through technological change, economic shifts, or organizational restructuring—you're left not just without a job, but without the resilience, adaptability, and diverse capabilities needed to thrive in a changed landscape.

The cruelest aspect of the security trap isn't that it fails to deliver on its promise. It's that it actively undermines the very qualities that create genuine resilience in an unpredictable world.

The Anti-Fragility Mindset: From Security to Resilience

SECURITY MINDSET	ANTI-FRAGILITY MINDSET
Optimizes for: Stability, predictability, comfort	Optimizes for: Growth, optionality, resilience
Career strategy: Long tenure at established organizations	Career strategy: Strategic moves based on learning and impact
Skill approach: Deep specialization in established domains	Skill approach: T-shaped expertise combining depth and breadth
Risk stance: Avoids uncertainty and change	Risk stance: Seeks calculated exposure to productive uncertainty
Identity source: Derives from job title and organization	Identity source: Derives from capabilities and impact
Network type: Deep connections in one industry/domain	Network type: Diverse connections across multiple domains
Financial approach: Maximizes stable income	Financial approach: Balances stable income with growth opportunities
Success measure: Advancement within predefined paths	Success measure: Expanded capability to create value in multiple contexts

Mark spent 11 years in middle management at a regional bank, turning down opportunities that seemed "risky" in favor of the predictable path. When his division was automated, he found himself competing for new positions against professionals who had developed diverse skill sets across multiple organizations.

"I realized I hadn't really gained 11 years of experience," Mark reflected. "I had gained 2-3 years of experience, repeated four times. Each time I chose security over growth, I was actually becoming less secure in the long run."

The opportunity cost of the security mindset extends far beyond missed career paths. It's about the skills not developed, the relationships not formed, the innovations not created, and the impact not achieved.

Research shows that early-career professionals who prioritize skill development and adaptability over immediate job security can achieve significantly higher long-term career outcomes.

A 2023 study from the Stanford Center on Longevity found that workers who pursued continuous learning and embraced job mobility in their 20s and 30s reported 25% higher job satisfaction and 18% higher earnings by mid-career compared to those who prioritized stability, a gap that widened over time (Carstensen et al., 2023). This isn't a minor edge—it reshapes career trajectories.

THE ANTI-FRAGILITY AUDIT: ASSESS YOUR CAREER VULNERABILITY

Before you can build career anti-fragility, you need to understand your current vulnerabilities. This audit will help you identify where you're most exposed to disruption and where you have the greatest opportunities to build resilience.

Rate yourself on a scale of 1-10 for each dimension below, where 1 represents high fragility and 10 represents high anti-fragility:

1. Income Diversification Score (__/10) How dependent are you on a single income source?

- Score 1-3: 90%+ of income from a single employer
- Score 4-6: Multiple income streams, but one source dominates (70%+)
- Score 7-10: Three or more significant income sources, none exceeding 50% of total

2. Skill Transferability Score (__/10) How valuable would your core skills be if your industry underwent major disruption?

- Score 1-3: Skills highly specific to current role/industry
- Score 4-6: Some skills are transferable, but would require significant retraining
- Score 7-10: Core skills valuable across multiple industries and contexts

3. Network Diversity Score (__/10) How diverse is your professional network across industries and functions?

- Score 1-3: Network concentrated in current company/industry
- Score 4-6: Some connections outside primary field, but limited depth
- Score 7-10: Strong relationships across multiple industries and functions

4. Financial Runway Score (__/10) How long could you sustain yourself if your primary income disappeared?

- Score 1-3: Less than 3 months of essential expenses
- Score 4-6: 3-6 months of runway

- Score 7-10: 6+ months of runway

5. Learning Velocity Score (__/10) How quickly can you acquire and apply new skills and knowledge?

- Score 1-3: Limited recent experience with significant learning
- Score 4-6: Occasionally learn new skills but not systematically
- Score 7-10: Regularly master new capabilities through deliberate practice

Total Anti-Fragility Score: __/50

- 40-50: Highly Anti-Fragile -- Well-positioned for disruption
- 25-39: Moderately Anti-Fragile -- Some vulnerabilities to address
- 10-24: Highly Fragile -- Immediate action needed to reduce vulnerability

Most professionals assume they're secure. But after taking this audit, you might realize—your career isn't resilient. It's just untested. That's a terrifying truth at first, but also a liberating one. Because once you see where you're fragile, you can start becoming anti-fragile.

The audit isn't just a measurement tool—it's a wake-up call. For many, it's the first time they've confronted the uncomfortable reality that their professional foundation might be built on shifting sand. But that recognition is also the first step toward building something stronger.

Action Steps Based on Your Score:

If you scored 10-24 (Highly Fragile):

- Prioritize building at least 3 months of financial runway immediately
- Begin developing a side project that creates a secondary income stream
- Join two professional communities outside your current industry
- Set aside 5 hours weekly for deliberate skill development

If you scored 25-39 (Moderately Anti-Fragile):

- Expand your highest-scoring area to further strengthen your position
- Focus on your two lowest-scoring areas for targeted improvement
- Create a systematic approach to developing transferable skills
- Build relationships with professionals in adjacent industries

If you scored 40-50 (Highly Anti-Fragile):

- Mentor others in building anti-fragility
- Develop systems that further enhance your strengths
- Consider entrepreneurial opportunities enabled by your anti-fragile position
- Explore emerging fields where your resilience creates competitive advantage

ANTI-FRAGILE CAREER CAPITAL: TURNING DISRUPTION INTO ADVANTAGE

Traditional career advice focuses on building what sociologist Pierre Bourdieu (1986) called "cultural capital"—the credentials, experiences, and relationships valued within established systems. While this remains important, the AI era demands a different form of professional asset: anti-fragile career capital.

Taleb's *Antifragile* (2012) argues that thriving under disorder requires adaptability over rigidity. Recent research confirms this in AI-driven job markets, where Bone et al. (2024) found a 30% rise in skills-based hiring for AI and green roles from 2020-2023, signaling an antifragile shift away from fragile degree reliance.

In the context of careers, an anti-fragile professional path is not just about surviving change but leveraging it as a catalyst for growth and reinvention.

Think of your anti-fragile career capital as an investment portfolio where:

- Skills = Assets that appreciate or depreciate in value over time
- Networks = Diversification that protects against single-point failures
- Experiences = Compound Interest that grows exponentially with each new challenge
- Opportunities = Liquidity that allows you to move quickly when conditions change

Like any sophisticated investor, your goal isn't to eliminate risk—it's to take intelligent risks that create asymmetric upside while limiting potential downside.

THE COMPONENTS OF ANTI-FRAGILE CAREER CAPITAL

1. Transferable skills that appreciate in value Not all skills are created equal in their resilience to disruption. Skills that apply across multiple domains and contexts create significantly more security than those tied to specific roles or technologies. High anti-fragility skills include:

- Strategic problem framing (defining the right problem)
- Systems thinking (understanding complex interactions)
- Stakeholder management (navigating human dynamics)
- First-principles reasoning (breaking problems into foundational elements)

Low anti-fragility skills include:

- Software-specific technical skills (tied to particular platforms)
- Process management for established workflows (vulnerable to automation)
- Middle-management coordination (being replaced by collaborative tools)

2. Diverse problem-solving experiences Each novel problem you solve creates a pattern-matching template in your professional repertoire. The more diverse these templates, the more adaptable you become to new challenges. Anti-fragile professionals deliberately seek problems:

- Across different industries and contexts
- With varying degrees of uncertainty and constraints
- That require different types of problem-solving approaches
- At different organizational levels and stages

3. Self-generated proof of impact When your value is demonstrated through tangible outcomes rather than role occupation, you become less dependent on institutional validation and more able to create opportunity regardless of formal position. Anti-fragile validation includes:

- Documented case studies of problems solved and value created
- Quantifiable metrics showing your specific impact
- Client/stakeholder testimonials that highlight your unique approach
- Intellectual property you've developed independently of employers

4. Knowledge networks across domains The most valuable professional connections span different industries, functions, and specialties, creating visibility into opportunities invisible within single-domain networks. Anti-fragile networks:

- Connect you to diverse expertise you can access when needed
- Provide early signals of disruption and opportunity
- Create multiple pathways to new roles and projects
- Offer stability independent of any single organization

THE INSECURITY ADVANTAGE: WHY EMBRACING UNCERTAINTY CREATES REAL SECURITY

Maya left a stable position at a top consulting firm to build her portfolio career—a combination of independent consulting, fractional leadership roles, and equity partnerships in early-stage ventures. Friends called her decision reckless, warning that she was sacrificing security for uncertainty.

Four years later, when economic contraction led to massive layoffs across consulting firms, Maya's diverse income streams and broad professional network provided far more resilience than her former colleagues' "secure" positions.

"I'm not immune to disruption," Maya explained. "But I've built a professional identity and income structure that can bend without breaking. When one area faces headwinds, others often present new opportunities."

Maya's story illustrates what I call the insecurity advantage—the counterintuitive truth that strategically embracing uncertainty creates more genuine security than clinging to stability.

The insecurity advantage works through three mechanisms:

1. Diversification: When you stop depending on a single source of income and professional identity, no single disruption can devastate your career.
2. Adaptation Velocity: Regular exposure to new challenges accelerates your ability to learn and pivot—the exact skills needed during major transitions.
3. Opportunity Visibility: Those comfortable with uncertainty develop what venture capitalists call "pattern recognition"—the ability to spot emerging opportunities before they become obvious to others.

THE ANTI-FRAGILE PLAYBOOK: FIVE STRATEGIES FOR THRIVING IN UNCERTAINTY

Strategy 1: Build Your Minimum Viable Security First

Anti-fragility doesn't mean recklessness. It means building a foundation of essential security that frees you to take calculated risks. Think of it as your career insurance policy.

Key actions:

- *Create financial runway:* Build 6-12 months of essential expenses in liquid savings
- *Develop a baseline of market-transferable skills:* Ensure you have capabilities that are valuable across multiple contexts
- *Establish your career documentation system:* Create a system for capturing your achievements, skills, and impact independent of any employer
- *Build your resilience infrastructure:* Develop practices that maintain your mental, physical, and emotional well-being amid uncertainty

Financial research consistently shows that having adequate emergency savings is one of the most important factors enabling professional risk-taking. A study published in the *Journal of Financial Planning* found that professionals with 6+ months of living expenses saved were significantly more likely to pursue entrepreneurial opportunities and navigate career transitions successfully (Grable & Joo, 2001), a finding echoed in recent data showing a 40% higher **likelihood of such moves among those with similar savings (Hanna et al., 2023).**

Strategy 2: Diversify Your Professional Portfolio

Anti-fragile professionals think like investors. They build diversified portfolios of skills, income streams, and opportunities rather than going "all in" on a single path.

Key actions:

- Skill diversification: Develop complementary skill sets that create unique combinations

- Income diversification: Build multiple revenue streams (employment, freelance work, passive income)
- Network diversification: Cultivate relationships across different industries, functions, and organizational types
- Opportunity diversification: Maintain multiple possible career paths you can activate as conditions change

"I don't have a five-year plan," explains marketing executive Sarah Jenkins. "I have a portfolio of skills, relationships, and directions I'm developing simultaneously. When disruption occurs, I don't have to start over—I just shift emphasis to the options that align with the new reality."

Strategy 3: Seek Asymmetric Upside

Anti-fragile professionals look for opportunities with asymmetric risk-reward profiles—where potential gains significantly outweigh potential losses.

Key actions:

- *Low-cost experiments:* Design small tests of new directions that provide valuable learning with minimal downside
- *Option creation:* Pursue opportunities that open multiple future paths rather than closing them
- *Skill acquisition leverage:* Identify capabilities that create disproportionate value across multiple contexts
- *Status game arbitrage:* Find domains where your existing expertise is rare and therefore more valuable

The concept of asymmetric upside is well-established in investment literature, but its application to career strategy is still emerging. Research by Harvard Business School professor Karim Lakhani found that professionals who engage in "low-cost experimentation" across multiple domains tend to discover unexpected career opportunities with disproportionate returns (Lakhani & Wolf, 2005), a pattern reinforced by recent experiments showing AI-assisted prototyping boosts productivity and quality by over 12% and 25%, respectively (Dell'Acqua et al., 2023).

Strategy 4: Embrace Strategic Job-Crafting

Rather than fitting yourself into predefined roles, anti-fragile professionals actively reshape their roles around their unique capabilities and emerging opportunities.

Key actions:

- *Value-gap identification:* Identify organizational needs not addressed by formal structures
- *Task negotiation:* Systematically shift your responsibilities toward high-impact, high-learning activities
- *Opportunity creation:* Propose new initiatives that leverage your unique skills while creating organizational value
- *Role redefinition:* Reframe your position around outcomes rather than activities

Rather than viewing job descriptions as rigid roles, the most resilient professionals treat them as flexible frameworks that evolve over time. Job crafting—proactively reshaping one's role to better align with strengths, interests, and organizational needs—has been extensively studied by organizational psychologists. Research published in the *Academy of Management Review* found that employees who engage in job crafting report higher levels of engagement, job satisfaction, and resilience during organizational change (Wrzesniewski & Dutton, 2001), a benefit reinforced in studies showing crafters adapt more effectively to workplace shifts, such as those driven by organizational change (Petrou et al., 2018).

Strategy 5: Build Your Personal Innovation Lab

Anti-fragile professionals create structured systems for continuous experimentation and learning outside their immediate responsibilities.

Key actions:

- Learning sprints: Dedicate focused periods to mastering specific capabilities
- Side projects: Develop ventures that test new skills and create alternative income streams
- Deliberate exposure: Systematically expose yourself to emerging trends and technologies
- Insight documentation: Create systems for capturing and synthesizing your learning

The most valuable professional asset isn't what you already know—it's your ability to quickly learn what you don't. This hinges on deliberate practice, a targeted method of skill-building through feedback and refinement. Pioneered by psychologist K. Anders Ericsson, it's proven to accelerate expertise across fields (Ericsson et al., 1993).

Case Study in Anti-Fragility: The Finance Professional Who Built Resilience

Michael had built a successful 12-year career in financial analysis at a major investment bank. His job seemed secure—until his entire department was restructured when AI systems dramatically reduced the need for human analysts performing routine market assessments.

Unlike many colleagues who were caught unprepared, Michael had been systematically building anti-fragile career capital for years:

1. Skill diversification: Beyond his core financial analysis expertise, Michael had developed capabilities in data science, strategic communication, and emerging blockchain technologies.
2. Network expansion: He had cultivated relationships across multiple industries through a finance podcast he created as a side project.
3. Impact documentation: He maintained detailed case studies of his most significant projects, quantifying his specific contributions independent of his formal role.
4. Options creation: He had conducted small consulting engagements that tested market demand for his expertise outside traditional banking.

When restructuring eliminated his position, Michael activated his anti-fragile assets. Within six weeks, he had three distinct opportunities:

- A fintech startup seeking his combined finance and technology expertise
- A consulting role leveraging his documented impact with similar clients
- A fractional CFO position with a company he'd met through his podcast

"What looked like a career disaster became a career catalyst," Michael explained. "The disruption forced me to fully leverage assets I'd been building but underutilizing. My income is now higher, my work more engaging, and most importantly, I'm no longer dependent on any single organization for my professional identity or economic security."

THE ANTI-FRAGILE MINDSET: PSYCHOLOGICAL FOUNDATIONS OF CAREER RESILIENCE

The transition from security-seeking to anti-fragility requires more than strategic actions—it demands fundamental shifts in how you think about work, value, and professional identity.

From Role Attachment to Capability Identity

When professional identity becomes fused with a specific role, organizational change becomes an existential threat rather than an adaptive challenge.

Anti-fragile professionals anchor their identity in capabilities and impact, not positions or titles. "I'm not a financial analyst who sometimes works on sustainability initiatives," explains David Chen. "I'm someone who combines financial and sustainability expertise to help organizations allocate capital more effectively. How that manifests in any given role is secondary to the value I create."

This shift creates psychological resilience during transitions. When you define yourself by what you can do rather than by your job title, organizational changes may affect your employment but not your fundamental professional identity.

From Certainty Seeking to Uncertainty Tolerance

The capacity to function effectively amid uncertainty is perhaps the most valuable professional skill in the AI era. Anti-fragile professionals deliberately develop their uncertainty muscle through regular exposure to ambiguous situations.

"I take on at least one project each year that terrifies me," shares consultant Rebecca Chen. "Not because I'm reckless, but because stretching my capacity to navigate uncertainty is the best insurance policy I can develop."

This practiced uncertainty tolerance creates competitive advantage. While others freeze during disruption, waiting for clarity before acting, those comfortable with ambiguity can move decisively amid incomplete information—often capturing opportunities invisible to certainty seekers.

From External Validation to Internal Metrics

The security mindset fosters dependency rather than agency. When you believe your fate lies in the hands of organizational decision-makers rather than your own choices, you develop what psychologists call an "external locus of control"—the belief that your outcomes are determined primarily by forces outside yourself.

Anti-fragile professionals develop clear internal metrics for progress and success, independent of external validation. "I track the expansion of my capabilities, the diversity of problems I can solve, and the tangible impact of my work," explains marketing strategist James Liu. "These metrics remain relevant regardless of my formal position or others' assessments."

This internal validation system creates stability amid organizational turmoil, allowing you to maintain confidence and direction even when external structures are in flux.

From Scarcity to Abundance Thinking

The security mindset often emerges from scarcity thinking—the belief that opportunities are limited and losses are permanent. This creates risk aversion that prevents growth and exploration.

Anti-fragile professionals cultivate abundance thinking—the recognition that disruption creates opportunities and that capabilities can be applied in countless contexts. "When my department was eliminated, my colleagues saw only what they'd lost," recalls finance professional Sarah Martinez. "I immediately began exploring how my skills could create value in new environments. While they were grieving their old roles, I was discovering possibilities they couldn't see."

This abundance orientation transforms how you experience change—from threat to opportunity, from ending to beginning, from loss to possibility.

Building Your Anti-Fragility Plan: A 12-Month Roadmap

Transforming your career from fragile to anti-fragile requires systematic action, not just conceptual understanding. Here's your actionable roadmap:

Month 1-3: Assess and Build Your Foundation

1. Complete your Anti-Fragility Audit: Identify your current vulnerabilities and resilience opportunities.

2. Create your financial runway plan: Develop a concrete strategy for building 6-12 months of essential expenses in liquid savings.
3. Establish your career documentation system: Implement tools and practices for systematically capturing achievements, skills, and impact.
4. Conduct your learning gap analysis: Identify the highest-leverage capabilities that would increase your professional mobility.

Month 4-6: Launch Your Anti-Fragile Experiments

1. Identify low-risk, high-learning opportunities: Design 2-3 small experiments that test new professional directions with minimal downside.
2. Initiate strategic skill development: Begin dedicated learning in areas that expand your opportunity landscape.
3. Activate network diversification: Strategically build relationships in domains adjacent to your current expertise.
4. Explore complementary income streams: Investigate options for creating value beyond your primary employment.

Month 7-12: Scale Your Anti-Fragile Assets

1. Implement your portfolio strategy: Formalize your approach to managing multiple professional dimensions simultaneously.

2. Expand your opportunity radar: Develop systems for identifying emerging possibilities before they become obvious to others.
3. Create your reinvention protocol: Establish triggers and processes for periodic reassessment and evolution of your professional focus.
4. Build your personal board of directors: Assemble diverse advisors who provide perspective and guidance across multiple domains.

The Liberating Truth: Embracing Productive Insecurity

The most liberating professional realization isn't finding perfect security—it's recognizing that such security never truly existed. What looked like stability was always just a temporary alignment between your capabilities and market demands. That alignment has always been subject to disruption, whether from technological change, economic shifts, or organizational transformation.

This reality isn't cause for anxiety—it's cause for empowerment. When you stop outsourcing your security to organizations or roles and start building it through your own capabilities and choices, you don't become more vulnerable—you become more resilient.

Research on career resilience demonstrates that professionals who acknowledge the inherent uncertainty of work are better equipped to navigate disruption. Stanford psychologist Carol Dweck's research on growth mindset reveals that individuals who see challenges as opportunities for development rather than threats to their identity are significantly more successful in navigating professional transitions (Dweck, 2006).

This mindset shift doesn't mean abandoning all structure or taking reckless risks. It means recognizing that true security was never about avoiding change—it was about building the capacity to thrive within it.

The professionals who thrive in the AI era won't be those clinging to disappearing certainties. They'll be those embracing the liberating truth that real security has always come from adaptability, not stability; from capabilities, not positions; from growth, not comfort.

ACTION BOX: TODAY'S FIRST STEP

Complete Your Anti-Fragility Quick Assessment

Take 5 minutes right now to honestly assess your current anti-fragility:

1. How many income sources do you have?
2. How many of your core skills would remain valuable if your industry changed dramatically?
3. How diverse is your professional network across different domains?
4. How many months could you sustain yourself if your primary income disappeared?

If your answers reveal vulnerability, don't wait for clarity. If you wait for disruption to hit, it's already too late. You need to act now. So ask yourself: What's one step you can take today to make yourself more resilient? Now take it. Because in a world that moves this fast, action isn't just a choice—it's survival.

The harshest truth about security is that it's always an illusion—until it's too late. Karen thought she was playing it safe. She wasn't. She was standing still while the world moved. The professionals who cling to stability don't see the cracks beneath their feet until they fall through. AI won't take their jobs—their own stagnation will. Anti-fragility isn't a choice between risk and safety. It's a choice between controlled growth or inevitable collapse. You don't need certainty to move forward. You just need the courage to stop standing still.

Now that you understand how to build career anti-fragility, it's time to turn these concepts into a practical playbook for thriving in the AI age. In Part 2, we'll explore exactly how to become an AI Architect—someone who designs their future rather than just adapting to it.

From Exposure to Execution

You've seen how AI is changing the landscape—and why outdated career playbooks are failing. But awareness alone isn't enough. Now, it's time to shift from defense to offense. The next section is about **taking control**, not just reacting. You don't need to be an AI expert to thrive—you just need to start seeing opportunities where others see obstacles. Let's reframe how you think about your role in an AI-driven world.

THREE THINGS FOR THIS WEEK

Reality Check: Stability Is a Trap—Resilience Is the Real Power Move

1. Rate yourself on adaptability: Are you actively learning or passively reacting? Identify one area to improve.

2. Set up a Horizon Scanning System—subscribe to one industry newsletter or AI trend report.
3. Test an AI tool in your workflow—not for efficiency, but to understand how it could change your role.

PART 2

HOW TO BECOME AN AI ARCHITECT INSTEAD OF AN AI VICTIM

A new way of thinking, working, and designing your AI-powered career

WHERE WE'RE HEADED

Now that we've dismantled the myths surrounding AI and work, it's time to shift our focus to action. The winners in the AI era aren't just using AI tools—they're designing the future. In this section, you'll learn how to stop thinking like an employee and start thinking like an AI Architect. We'll explore how to redefine your value, stack your skills strategically, and leverage AI as an amplifier rather than a threat. By the end of this section, you'll have the mindset and tactics needed to stay ahead.

05

STOP THINKING LIKE AN EMPLOYEE
START ACTING LIKE AN AI ARCHITECT

Employees wait. AI Architects create.

By 2027, there will be only two types of knowledge workers left: those who direct AI systems and those who take orders from them.

If Part 1 exposes the broken rules of career success, Part 2 is about rewriting them. This chapter will show you how to shift from reacting to AI-driven changes to designing your career with AI as your strategic advantage.

Jason, a data analyst at a financial services firm, had just witnessed his first true AI breakthrough moment. The dashboard that normally took him days to compile—gathering data, cleaning it, creating visualizations, adding annotations—had been assembled by an AI system in less than an hour. And if he was being honest, the AI's version was better.

"My first thought was pure panic," Jason told me. "If my main value is creating these reports, and AI can do them faster and better, what am I even doing here?"

But instead of resisting the change, Jason pivoted. **He stopped seeing himself as a report creator and started seeing himself as an insight architect.** He realized AI could handle the execution, but it couldn't frame the right business questions, synthesize insights into strategic decisions, or understand the nuances of stakeholder needs.

Within six months, Jason's role had transformed—**from data analyst to strategic advisor.**

This is the mindset shift this chapter is about: **Stop thinking like an employee. Start thinking like an AI Architect.**

The difference isn't semantic—it's existential. Employees follow instructions, execute tasks, and work within defined boundaries. They are increasingly vulnerable as AI automates routine execution. AI Architects, in contrast, design workflows, orchestrate intelligence systems, and create value at a level of abstraction above mere task completion.

This shift isn't about eliminating jobs—it's about evolving how work creates value. Those who see this change as an opportunity will find themselves in higher demand than ever before—not because they resist AI, but because they learn to lead it.

THE EXTINCTION OF THE EMPLOYEE MINDSET

The traditional employee is becoming obsolete faster than anyone wants to admit.

For decades, the fundamental equation of employment has been deceptively simple: perform assigned tasks competently = receive compensation and career advancement. This equation created a mindset focused on executing pre-defined work within established parameters. Success meant following directions effectively, meeting expectations reliably, and gradually expanding responsibilities through demonstrated competence.

This model is collapsing before our eyes. AI systems can now execute an expanding range of knowledge work with greater speed, consistency, and often higher quality than humans. What's more, these systems improve exponentially while humans improve incrementally.

Consider these sobering realities:

- Language models can now draft reports, emails, presentations, and code at a level matching or exceeding average professional standards
- Data analysis tools can automate the creation of insights from raw information with minimal human guidance
- Design systems can generate visual content based on text prompts that rival mid-level creative output
- Project management tools can now optimize resource allocation and timeline management better than most human managers

The death of the employee mindset isn't coming—it's already here. Those who cling to the old equation of "task execution =

value" are already experiencing diminishing returns on their professional investment.

The warning signs of the employee mindset aren't subtle. You're operating with this increasingly dangerous orientation if:

- You define your value primarily through tasks you complete rather than outcomes you enable
- You wait for formal authorization before addressing problems you've identified
- You see your role as fixed rather than continuously evolving
- You measure success by meeting expectations rather than creating new possibilities
- You wait for the organization to provide learning opportunities rather than self-directing your development

If three or more of these descriptions sound familiar, you're operating with an employee mindset that creates significant vulnerability in the AI era. The good news? This mindset isn't fixed. It can be transformed through deliberate reframing of your professional identity and approach.

THE AI ARCHITECT BLUEPRINT: YOUR NEW PROFESSIONAL IDENTITY

The AI Architect doesn't just use AI tools—they design the systems in which human and artificial intelligence combine to create unprecedented value.

This isn't merely a job title—it's a fundamentally different approach to professional value creation. Rather than defining themselves by the tasks they perform, AI Architects see themselves as designers of systems that create value, with or without their direct execution.

"The shift is from 'I create value by doing things' to 'I create value by designing how things get done,'" explains organizational psychologist Dr. Amy Edmonson. "This doesn't mean you never execute tasks. It means your primary value comes from architecting approaches rather than handling execution" (Edmonson, 2019).

This mindset shift parallels the evolution of software development. Traditional programmers focused on writing code that executed specific functions. Modern developers increasingly work as architects who design systems, leveraging existing components and platforms to create solutions at a higher level of abstraction.

THE THREE CORE RESPONSIBILITIES OF THE AI ARCHITECT

Whether you're a marketing specialist, financial analyst, project manager, or any other knowledge worker, the AI Architect approach centers on three core responsibilities:

1. Design Intelligent Workflows

The AI Architect analyzes end-to-end value creation processes and designs optimal combinations of human and machine capabilities. This includes:

- Mapping the current state of work processes
- Identifying where AI can handle routine execution

- Determining where human judgment adds unique value
- Creating feedback mechanisms that improve both human and AI performance over time

2. Orchestrate Collaborative Intelligence

The AI Architect conducts the symphony of human and artificial intelligence, ensuring each contributes optimally to desired outcomes. This includes:

- Developing frameworks for human-AI collaboration
- Creating effective prompts and instructions for AI systems
- Establishing appropriate human oversight and intervention points
- Designing interfaces between AI outputs and human decision-making

3. Translate Between Domains

The AI Architect bridges technical, business, and human dimensions to create integrated solutions. This includes:

- Translating business needs into technical specifications for AI systems
- Converting AI outputs into actionable business insights
- Communicating complex technical concepts to non-technical stakeholders
- Ensuring AI implementations address genuine human and business needs

THE AI ARCHITECT IN ACTION: BEFORE AND AFTER

Let's see how the AI Architect approach transforms work across different domains:

Marketing:

- **Employee Approach**: Creates content, manages campaigns, analyzes performance metrics
- **AI Architect Approach**: Designs content strategy that leverages AI for creation while focusing human attention on brand voice alignment; orchestrates AI-powered campaign optimization while adding human understanding of customer psychology; creates frameworks that translate AI-generated analytics into strategic decisions

Finance:

- **Employee Approach**: Builds financial models, generates reports, conducts analysis
- **AI Architect Approach**: Designs intelligent systems that automatically generate financial insights; orchestrates collaboration between AI forecasting tools and human strategic judgment; translates complex financial data into decision frameworks for non-financial leaders

Product Development:

- **Employee Approach**: Executes design and development tasks according to specifications

- **AI Architect Approach**: Creates frameworks for AI-augmented product innovation; orchestrates systems where AI handles routine coding while humans focus on user experience and strategic direction; translates between technical capabilities, market needs, and business strategy

The pattern is consistent across domains—the AI Architect operates at a higher level of abstraction, designing how work gets done rather than simply doing assigned work.

THE AI-PROOF FRAMEWORK: HOW TO STAY AHEAD IN AN AI-DRIVEN WORLD

To thrive in the AI era, you need a structured approach to transforming from employee to Architect. The AI-Proof Framework provides this roadmap through three essential components:

1. Reposition Your Value: From Employee To Architect

Before AI: Employees define themselves by their tasks. **After AI**: AI Architects define themselves by the problems they solve and the value they create.

Maya had spent eight years as a data analyst at a healthcare company when generative AI began transforming her field. While her colleagues panicked or ignored the changes, Maya took a more strategic approach.

"I realized I needed to fundamentally redefine how I created value," Maya explained. "Before AI, I defined myself by the reports

I produced. After AI, I needed to define myself by the business problems I solved."

Maya began a systematic process of repositioning her value through three key steps:

First, she tackled **Identity Reframing**—shifting how she thought about her professional role. "I started introducing myself as a 'healthcare insights architect' rather than a 'data analyst,'" she recalled. "This wasn't just semantics. It changed how I approached every aspect of my work."

Next came **Value Articulation**—redefining her value proposition based on outcomes rather than activities. "I rewrote my internal documentation to focus on how my work improved patient outcomes and operational efficiency, not just what reports I delivered," Maya explained.

Finally, she implemented **Work Restructuring**—reorganizing her activities around design, orchestration, and translation. "I spent less time creating visualizations and more time designing workflows where AI could generate insights that I could then contextualize for stakeholders."

The transformation wasn't immediate, but the results were dramatic. When her company restructured analytics operations six months later, Maya wasn't just retained—she was promoted to lead a new insights team with AI at its core.

THE ARCHITECT ADVANTAGE: WHAT AI CAN'T REPLACE

Understanding the division of labor between human and machine intelligence was fundamental to Maya's success. She created a simple mapping that guided her transformation:

"For every task where AI excelled—like recognizing patterns in massive datasets—I looked for the complementary human capability—like spotting novel connections between seemingly unrelated domains," Maya explained. "Instead of competing with AI on its strengths, I doubled down on distinctly human capabilities."

This approach created a powerful complementary relationship between Maya and AI systems. While AI handled prediction based on historical data, Maya focused on judgment in ambiguity. While AI generated multiple variations quickly, Maya concentrated on determining which variations actually mattered and why.

"The breakthrough came when I stopped seeing this as a competition and started seeing it as a partnership," Maya reflected. "AI was handling what machines do best, freeing me to focus on what humans do best."

WHAT AI DOES BEST	WHAT ARCHITECTS DO BEST
High-Volume Pattern Recognition Spotting regularities across massive datasets	Novel Pattern Recognition Identifying connections between seemingly unrelated domains
Prediction Based on Historical Data Forecasting outcomes using existing patterns	Judgment in Ambiguity Making decisions with incomplete information and unclear outcomes
Rapid Iteration Generating multiple variations quickly	Value-Based Discernment Determining which variations actually matter and why
Rules-Based Decision Making Applying consistent logic at scale	Contextual Understanding Interpreting situations considering cultural, ethical and human factors
Information Processing Analyzing structured data efficiently	Meaning-Making Creating narratives that build meaning from information
Answering Defined Questions Responding to clear queries	Asking Essential Questions Identifying which questions should be asked in the first place
Optimizing Known Metrics Maximizing defined success measures	Defining Success Metrics Determining what should be measured and why

The AI Architect doesn't compete with AI on the left column—that's a losing battle. They focus on mastering the right column, then strategically direct AI to handle the left.

When Carlos, a marketing specialist at a technology firm, first attempted his value repositioning audit, he struggled. "I kept listing my job responsibilities instead of thinking about outcomes," he admitted. After several iterations, he developed a more transformative perspective:

"I realized that creating monthly performance reports wasn't my real value," Carlos explained. "My actual contribution was enabling data-driven marketing strategy decisions. Writing campaign copy wasn't my value—driving customer engagement and conversion was. Processing analytics wasn't my value—ensuring performance optimization and ROI was."

This shift in thinking led Carlos to examine the broader systems his work supported. For the marketing strategy system, he identified stakeholders from sales, product teams, and executive leadership. He mapped inputs (market research, customer data, competitive analysis), processes (strategy development, content creation, channel optimization), and outputs (campaigns, customer engagement, revenue generation).

"Once I understood my place in these systems, I could spot opportunities to create more value through better design," Carlos explained. "For our content development system, I realized we could integrate AI content generation with human brand voice refinement to produce 3x the output with better results."

Carlos implemented this redesigned workflow the following week, focusing on orchestrating the complementary strengths of human creativity and AI productivity. "This wasn't about using AI tools occasionally—it was about reimagining our entire content creation system with AI as a core component and humans focusing on strategy and refinement."

2. Amplify With Ai: Tool, Not Threat

AI isn't coming for your job—it's coming for the tasks within your job. Your choice is whether to cling to those tasks or elevate above them.

The second component of the AI-Proof Framework focuses on using AI as a force multiplier rather than seeing it as a competitor or threat.

Michael's transformation as a content marketer illustrates the power of the force multiplier mindset. When generative AI first emerged, Michael noticed his colleagues taking two opposing approaches.

"Some ignored AI completely, insisting that 'real creativity' couldn't be replicated," Michael observed. "Others surrendered to it, basically using whatever the AI produced with minimal oversight. Neither approach was working well."

Michael developed a third approach: seeing AI as a force multiplier for his existing capabilities.

Before embracing this mindset, Michael's typical week was grueling. "I'd spend 30 hours writing first drafts, produce maybe 5 finished pieces, and have almost no time for strategy or optimization," he recalled. "I was constantly burned out and anxious about keeping up with volume demands."

After implementing his AI amplification strategy, everything changed. "Now I spend 5 hours directing AI to generate first drafts, which lets me produce 15 finished pieces weekly—triple my previous output," Michael explained. "More importantly, I've repurposed that time to invest 10 hours in audience research and content strategy, and another 10 hours in distribution optimization and performance analysis."

The transformation wasn't about working less—it was about creating significantly more value with the same hours. "I'm not working fewer hours," Michael emphasized. "I'm creating dramatically more value by letting AI handle production while I focus on direction. My value isn't in writing words—it's in determining which words will drive business results and why."

This approach transformed Michael's relationship with AI from competitor to collaborator. Instead of seeing AI as a threat to his creative identity, he recognized it as a tool that could handle routine execution while he focused on higher-level strategic work.

Michael's transformation was guided by four core principles:

"First, I focused on direction over execution—spending my energy directing AI systems rather than competing with them on routine tasks," Michael explained. "Second, I emphasized judgment over production—concentrating on making value judgments about what should be created rather than doing all the creation myself."

"Third, I invested in design over delivery—using my time to design approaches and frameworks rather than just delivering outputs. And finally, I prioritized integration over isolation—treating AI as an integrated part of my workflow rather than a separate tool I occasionally used."

The results spoke for themselves. Not only did Michael's productivity increase dramatically, but his job satisfaction improved as well. "I'm doing more meaningful work now," he reflected. "Instead of rushing through content creation, I'm actually thinking strategically about audience needs and business outcomes."

AI as a Superagency Enabler

The true AI winners aren't those who use it as a crutch but those who use it as an amplifier. McKinsey's research calls this *superagency*—where AI enhances an individual's ability to create, think, and act at a level previously impossible (Mayer et al., 2025). This means AI isn't just a tool; it's a power boost for those who know how to wield it. This concept reinforces that AI is fundamentally about augmenting human potential rather than replacing it—allowing professionals to operate with greater speed, creativity, and impact than ever before.

The AI Amplification Cycle: A Virtuous Loop

The most successful AI Architects implement a continuous cycle that maximizes the complementary strengths of human and machine intelligence:

1. **Frame**: Define the problem, context, and desired outcome with human judgment
2. **Direct**: Provide clear instructions and parameters to AI systems
3. **Generate**: Let AI create multiple potential solutions or outputs
4. **Evaluate**: Apply human judgment to assess and select the most valuable outputs
5. **Refine**: Direct AI to modify and improve selected outputs
6. **Integrate**: Incorporate the results into broader workflows and systems
7. **Learn**: Analyze the entire process to improve future cycles

This cycle creates a virtuous loop where AI handles volume and variation while humans provide direction, judgment, and integration.

ACTION STEP: CREATE YOUR AI AMPLIFICATION PLAN

To begin amplifying your strengths with AI rather than being replaced by it, follow these steps:

1. Task Inventory and Classification List all your regular work tasks and classify each as:

- **AI Advantage**: Tasks where AI can outperform humans (repetitive, pattern-based, data-intensive)
- **Human Advantage**: Tasks where humans outperform AI (judgment-intensive, contextual, novel, interpersonal)
- **Collaboration Advantage**: Tasks where human-AI partnership creates optimal results

2. Amplification Opportunity Identification For each task category, determine your approach:

- **AI Advantage Tasks**: How could you design better workflows that leverage AI for these tasks while adding human oversight?
- **Human Advantage Tasks**: How could AI tools free up more of your time and attention for these high-value activities?
- **Collaboration Advantage Tasks**: How could you design optimal human-AI workflows that leverage the strengths of both?

3. Tool Selection and Integration Identify 2-3 specific AI tools or capabilities that could help you implement your amplification strategy, and develop a concrete plan for implementing them in your workflow within the next 30 days.

3. Execute For Impact: The Race Is Already Started

While you're contemplating whether to transform, others are already gaining irreplaceable experience. The window to lead isn't indefinite.

Thomas and Emma both worked as financial analysts at competing investment firms. Both witnessed the same AI revolution in their field. Their responses couldn't have been more different.

Thomas took a wait-and-see approach. "I'll adapt when the technology stabilizes," he reasoned. "No need to jump on every new trend." He continued working as he always had, watching AI developments with casual interest but taking no immediate action.

Emma immediately began experimenting with AI tools, integrating them into her workflow, and reimagining her role as an insight architect rather than an analysis producer. "I don't have time to wait," she explained to colleagues. "By the time this becomes mandatory, it'll be too late to lead—you'll be struggling just to catch up."

Eighteen months later, the divergence was stark:

Thomas found himself increasingly marginalized as AI systems absorbed more of his routine tasks. His value proposition was shrinking, and his unique contributions were increasingly difficult to articulate. When his firm went through a restructuring, his position was among those eliminated.

Emma had transformed into an AI Architect who designed analytical approaches that leveraged both machine capabilities and human judgment. She'd been promoted twice and was now leading her firm's AI integration initiative across the investment division, with compensation more than double her starting salary.

The difference wasn't their starting skills, credentials, or positions. It was their response time and willingness to act decisively amid ambiguity.

The First-Mover Advantage in AI Integration

A growing body of evidence suggests that early AI adopters gain significant competitive edges. According to a February 2025 Forbes Technology Council article (Mewawalla, 2025), enterprise AI adoption has reached a tipping point, with early movers shifting from pilots to revenue-generating implementations. Citing IDC, the article notes global AI spending is projected to reach $632 billion by 2028, growing at 29% annually, driven by pioneers leveraging cost declines—like model costs dropping from $60 to $0.06 per million tokens—to optimize operations ahead of competitors.

This "first-mover advantage" comes from several sources:

1. **Compounding Learning**: Early experience with AI tools creates knowledge that compounds over time, allowing rapid adaptation to new capabilities
2. **Workflow Redesign**: First-movers can reimagine entire workflows around AI capabilities rather than simply plugging AI into existing processes

3. **Organizational Influence**: Early adopters often shape how their organizations implement AI, creating advantages in status and decision-making influence
4. **Relationship Building**: Those who pioneer AI adoption build relationships with technical teams and vendors that provide ongoing advantages in access and support

The window to capitalize on these advantages isn't indefinite. As AI adoption accelerates, the distinction between early and late adopters will clarify—with significant implications for career trajectory.

From Concept to Action: The 5 Daily Practices

The transformation from employee to Architect requires more than conceptual understanding. It demands concrete shifts in daily professional practice. Here are five daily habits that accelerate the transformation:

1. **Question-Centered Planning**: Rather than organizing your day around tasks to complete, structure it around questions to answer and problems to solve. This single shift immediately elevates your thinking from execution to architecture. **Practice**: Begin each day by identifying the 2-3 most valuable questions you could help answer, regardless of whether answering them falls within your formal responsibilities.
2. **Outcome Visualization**: Regularly visualize the ultimate outcomes your work enables, independent of the specific tasks involved. This reinforces outcome orientation over task orientation. **Practice**: Before starting any significant

work activity, explicitly identify the business or user outcome it supports, not just the task it completes.

3. **System Mapping**: Develop the habit of mapping the systems you participate in—identifying inputs, processes, outputs, and interactions across components. This builds your capacity to design and optimize systems rather than just operating within them. **Practice**: Once weekly, diagram a process you're involved in, identifying points where human judgment adds unique value versus where automation could enhance outcomes.

4. **Cross-Boundary Exploration**: Regularly investigate adjacent domains that intersect with your core expertise. This builds combinatorial knowledge that creates unique value. **Practice**: Dedicate 10% of your professional development time to learning about domains adjacent to your core expertise—not to become an expert, but to identify valuable connections.

5. **Value Articulation**: Practice explicitly articulating the value you create in outcome terms rather than task terms. This reinforces your identity as a value creator rather than a task performer. **Practice**: Rewrite your professional introduction or elevator pitch to focus entirely on the outcomes you enable rather than the tasks you perform.

These practices might seem subtle, but their cumulative impact is profound. They gradually rewire how you perceive your professional identity and value, shifting from task execution to system design.

THE AI ARCHITECT'S TOOLKIT: ESSENTIAL SKILLS FOR THE NEW ECONOMY

Beyond mindset and approach, specific skills distinguish the AI Architect from the traditional employee. These capabilities form the essential toolkit for thriving in the age of intelligent machines.

1. System Thinking

The ability to understand how components interact within complex systems is foundational to the Architect approach. System thinking involves:

- Identifying relationships and dependencies between parts
- Recognizing feedback loops and amplification effects
- Understanding how changes in one area affect the whole
- Anticipating second and third-order consequences

Development Strategy: Choose a complex process you're involved with and map all its components, connections, and feedback mechanisms. Identify points where AI could enhance system performance and where human oversight remains essential.

2. Prompt Engineering

As AI systems become more powerful, the ability to effectively direct them through well-crafted prompts becomes increasingly valuable. Effective prompt engineering includes:

- Framing clear, specific instructions
- Providing appropriate context and constraints
- Establishing evaluation criteria for outputs
- Designing iterative refinement processes

Development Strategy: Select an AI writing or image generation tool and experiment with systematically varied prompts. Document what works, what doesn't, and why. Create a personal library of effective prompt patterns for different purposes.

3. Strategic Translation

The capacity to translate between business needs, technical capabilities, and human factors is a core AI Architect skill. Strategic translation involves:

- Converting business objectives into technical requirements
- Translating technical outputs into business insights
- Connecting stakeholder needs to system designs
- Communicating complex concepts to diverse audiences

Development Strategy: Practice explaining a technical concept to three different audiences: technical experts, business leaders, and end users. Note how your explanations differ and what makes each effective for its audience.

4. Workflow Design

Creating optimal processes that combine human and machine capabilities is essential for maximizing value creation. Effective workflow design includes:

- Mapping current processes and identifying inefficiencies
- Determining appropriate human-AI division of labor
- Creating seamless handoffs between human and machine components
- Designing feedback mechanisms that improve both human and AI performance

Development Strategy: Select a workflow you're involved with and redesign it from first principles, assuming access to current AI capabilities. Compare your design to the existing process and identify key differences.

5. Ethical Oversight

As AI systems make more consequential decisions, the ability to provide ethical guidance and oversight becomes critical. Ethical oversight involves:

- Identifying potential biases in AI systems
- Considering diverse stakeholder perspectives and impacts
- Designing appropriate human intervention points
- Establishing clear accountability frameworks

Development Strategy: Select an AI application in your field and conduct an ethical impact assessment. Identify potential risks, affected stakeholders, and appropriate oversight mechanisms.

THE AI ARCHITECT'S CAREER ACCELERATOR: YOUR 90-DAY TRANSFORMATION PLAN

Elena's journey from traditional employee to AI Architect illustrates how theory translates into practice through deliberate action. As a product manager facing AI disruption in her industry, Elena developed a systematic 90-day plan to transform her approach.

First 30 Days: Foundation Building

Elena began her transformation with a fundamental mindset shift. "During week one, I completed my Value Repositioning Audit, which was eye-opening," she recalled. "I discovered that nearly 40% of my tasks could be automated, but the outcomes those tasks supported were still critically important."

She also started practicing question-centered planning, beginning each day by identifying the 2-3 most valuable questions she could help answer. "This simple change immediately elevated my thinking," Elena explained. "Instead of focusing on my task list, I started thinking about the problems worth solving."

By week two, Elena had mapped her current workflows and identified where AI could make the biggest impact. "I discovered that our customer feedback analysis process was a prime candidate for augmentation," she noted. "AI could process the raw feedback data while I focused on extracting strategic insights and implementation priorities."

During week three, Elena began experimenting with AI tools relevant to product management. "I started with a simple AI writing assistant for product documentation," she explained. "I wasn't looking to replace our documentation process but to understand how AI could enhance it."

The final week of her foundation phase focused on redesigning one key workflow. "I created a new system for our competitive analysis process that combined AI market research with human strategic interpretation," Elena shared. "This initial redesign created immediate value while building my confidence for bigger changes."

Days 31-60: Skill Development

The second month of Elena's transformation focused on developing specific AI Architect capabilities. "I spent weeks five and six becoming proficient at prompt engineering," she explained. "I developed templates for recurring tasks like feature descriptions and learned how different prompt structures produced dramatically different results."

This experimentation led to a personalized prompt library that Elena could leverage for consistent results. "By the end of week six, I had standardized prompts for ten different product management tasks, from user story generation to technical specification drafts," she noted.

Weeks seven and eight focused on refining her initial workflow redesigns. "As I gained experience, I spotted optimization opportunities in my competitive analysis process," Elena said. "I adjusted the handoffs between AI research and human interpretation, creating clearer criteria for what required my attention versus what could be delegated."

The final weeks of this phase involved communicating her evolving approach to colleagues. "I created documentation explaining how our team could leverage AI while maintaining our product quality standards," Elena recalled. "This wasn't just about sharing techniques—it was about helping others understand the strategic shift in how we worked."

Days 61-90: Impact Amplification

In the final month, Elena scaled her approach across additional workflows. "By week eleven, I had integrated AI capabilities into our product roadmapping, customer feedback analysis, and market research processes," she explained. "The systems were designed to enhance human judgment, not replace it."

She also developed feedback mechanisms to track results and drive improvement. "We created metrics comparing our AI-augmented output quality, consistency, and speed against our previous approaches," Elena noted. "This data-driven evaluation was crucial for continuous optimization."

By the end of the 90 days, Elena had transformed from a traditional product manager to an AI Architect who designed systems leveraging both human and artificial intelligence. "The final week was about documenting my journey and sharing these learnings with the broader organization," Elena reflected. "What started as a personal transformation evolved into a template that others could follow."

Elena's methodical approach demonstrates how commitment to systematic practice transforms theoretical understanding into practical results. "The key wasn't learning specific AI tools—it was developing a fundamentally different approach to creating value," she concluded. "By the end of 90 days, I wasn't just using AI—I was thinking like an AI Architect."

THE FUTURE BELONGS TO ARCHITECTS, NOT EMPLOYEES

By 2030, most jobs won't be replaced by AI—they'll be redesigned around it. Those who master the Architecture will lead this redesign. The rest will merely implement it.

The shift from employee to AI Architect isn't about abandoning your current expertise or seeking a completely different role. It's about fundamentally reimagining how you create value with the knowledge and capabilities you already possess.

This transformation occurs through deliberate evolution across five dimensions:

1. **Identity**: From "I am my role" to "I am the value I create" Begin by consciously reframing how you think about your professional identity. Instead of defining yourself by your title or function, define yourself by the outcomes you enable and the problems you solve.

2. **Focus**: From "tasks to complete" to "systems to design" Shift where you direct your attention and energy—from executing pre-defined activities to designing better approaches for value creation, with or without your direct execution.

3. **Value Creation**: From "doing assigned work" to "solving valuable problems" Reorient your contribution from completing assigned responsibilities to identifying and addressing problems that matter, regardless of whether they fall within your formal remit.

4. **Expertise Development**: From "getting better at what you do" to "expanding what you can do" Shift your learning focus from deepening existing capabilities to developing new capabilities that create unique combinatorial value, particularly at the intersection of domains.

5. **Relationship to Change**: From "adapting when necessary" to "driving transformation proactively" Transform your stance toward organizational and technological change from reactive adaptation to proactive leadership.

The age of the traditional employee is ending. What replaces it isn't unemployment—it's a new class of professional who designs, directs, and orchestrates how work gets done rather than simply doing assigned work.

The choice is stark but simple: Will you be the Architect designing tomorrow's work, or the employee desperately clinging to yesterday's tasks?

Those who make this transition proactively will find themselves leading the next era of professional value creation. Those who resist will find themselves increasingly marginalized as more of their tasks become automatable and their value proposition erodes.

The AI revolution isn't primarily about technology. It's about a fundamental reimagining of how humans create value—from execution excellence to architectural innovation. The machines aren't coming for your job. They're coming to transform how your job creates value.

Reframing your role is the first step. But to truly future-proof your career, you need a mindset that thrives in uncertainty. That's where curiosity becomes your greatest asset. Let's dive into why.

Now that we've dismantled the illusion of career stability, the real question isn't what AI will do to your job—it's what you will do with AI. You can't future-proof a role that was designed for the past, but you can build skills that make you indispensable in the AI-powered future. The professionals who thrive won't just work differently— they'll construct entirely new blueprints for value. Let's talk about how.

You don't have to fear AI. You have to **lead it**. The future doesn't belong to those who cling to tasks—it belongs to those who

orchestrate intelligence. You're not just adapting to AI. **You're designing the future of work itself.**

But mastering the AI Architect mindset is just the first step. The real superpower isn't just designing systems—it's asking the questions no one else is asking. The difference between those who adapt to AI and those who define the future will come down to one thing: curiosity.

THREE THINGS FOR THIS WEEK

Reality Check: If AI Can Measure It, AI Can Replace It

1. Look at your last three big work decisions—where did **human judgment** play a role AI couldn't replicate?
2. Identify **one uniquely human skill** that makes you valuable—how can you double down on it?
3. Find an AI-powered tool that's reshaping your industry and **learn its strengths and weaknesses**.

06

CURIOSITY ISN'T A SOFT SKILL
IT'S THE ULTIMATE EDGE

The most powerful skill isn't coding—it's curiosity.

In an AI-dominated world, your ability to ask brilliant questions will be worth 10x more than your ability to have all the answers.

The best AI architects don't just design—they discover. Curiosity isn't a luxury in the AI era; it's the engine that fuels reinvention.

Thomas had spent twenty years mastering his craft as a financial analyst. He'd earned the right credentials, climbed the corporate ladder, and built his reputation on being the person with answers. His expertise was his identity.

Then in 2023, his company deployed an AI financial analysis system that could process reports, identify patterns, and generate insights in seconds. Work that had taken Thomas days now happened instantaneously. Colleagues who once came to his office seeking wisdom now consulted the AI first.

"I've become obsolete," he confided to me during a consultation. "Everything I know, this thing knows better."

But Thomas was asking the wrong question. The real question wasn't whether AI knew more than him—it was whether he could explore better than AI.

Six months later, Thomas wasn't competing with AI—he was commanding it. While his colleagues used AI to answer standard questions, Thomas was using it to explore unconventional connections between market volatility and climate policy, designing new risk assessment models no one had thought to build. His value wasn't in having answers—it was in asking questions no one else was asking.

Success in an AI-powered world is not about knowing more— it's about exploring better.

This is the new intelligence hierarchy: AI can know, but only humans can wonder. AI can answer, but only humans can question. AI can optimize existing patterns, but only the curious can imagine new ones.

Curiosity isn't some fuzzy "nice-to-have" personality trait. It's not the soft skill that HR departments list as an afterthought. It's the fundamental driver of human advantage in the age of artificial intelligence—and it's completely AI-proof.

The moment AI exposes a pattern in your thinking, it also reveals an opportunity: to explore beyond it. Each time an algorithm replicates what you do, it's not eliminating your job—it's eliminating your excuses for not evolving. This is why curiosity isn't just another professional virtue—it's the ultimate AI-proofing mechanism.

CURIOSITY: THE #1 CAREER DIFFERENTIATOR IN THE AI ERA

Let's be brutally clear: **most professional skills you've developed throughout your career are already being automated.** Technical expertise, analytical abilities, content creation, coding, design—AI systems are rapidly matching or exceeding human performance in these domains.

What remains uniquely human isn't the ability to know more or work faster—it's the ability to wonder, to question, to imagine what doesn't yet exist. In the AI era, curiosity isn't just another professional asset; it's becoming the primary differentiator between those who will lead and those who will follow.

Research from Harvard Business School demonstrates that organizations with leaders who foster curiosity see substantial advantages in innovation, adaptation to market changes, and team performance (Gino, 2018). Yet most workplaces systematically undervalue curiosity, treating it as a personality quirk rather than a strategic advantage.

This misclassification creates enormous opportunity for those who recognize that curiosity isn't just a nice-to-have trait—it's a deliberate practice that can be developed, directed, and deployed for competitive advantage.

THE SCIENCE OF CURIOSITY: WHY IT MATTERS MORE NOW

Let's cut through the self-help fluff and understand what curiosity actually is—and why it matters more now than ever before.

The Neuroscience of Exploration

Psychologist Daniel Berlyne distinguished between two types of curiosity: diversive and epistemic. Diversive curiosity is the broad desire for novelty—the kind that has you scrolling through social media or channel-surfing. It's shallow and fleeting. Epistemic curiosity, however, is the deep drive to understand—to close specific knowledge gaps and solve meaningful problems. It's epistemic curiosity that drives innovation, expertise, and career resilience (Berlyne, 1960).

George Loewenstein's (1994) "information gap theory" explains why curiosity is so powerful: our brains are wired to notice gaps between what we know and what we want to know. When we become aware of these gaps, it triggers a form of cognitive discomfort that motivates us to seek answers. This isn't just a personality trait—it's a neurological drive that can be developed and directed.

The professional impact of this drive is profound. Research at George Mason University has shown that curious professionals are more likely to generate creative solutions, adapt to changing circumstances, and build diverse knowledge networks. They don't just perform better—they evolve faster (Kashdan & Silvia, 2009).

But here's where most people get it wrong: they treat curiosity as an inherent trait rather than a strategic skill. They believe some people are "naturally curious" while others aren't. That's nonsense. Curiosity is a muscle that can be strengthened, a habit that can be cultivated, and a competitive advantage that can be weaponized.

The Neurobiological Rewards of Curiosity

The science goes deeper. Neuroimaging studies reveal that curiosity activates the brain's reward pathways—the same systems involved in anticipating positive outcomes. When we're curious, the brain releases dopamine, the neurotransmitter associated with motivation and pleasure. This creates a self-reinforcing loop: curiosity leads to discovery, discovery triggers reward, and reward strengthens curiosity (Gruber et al., 2014).

This explains why curious professionals report higher job satisfaction and are less vulnerable to burnout. They're literally getting a neurochemical reward for exploring. While their colleagues experience work as a series of obligations, curious professionals experience it as a series of discoveries.

What's particularly interesting is that the brain shows greater activity in memory formation when we're curious. A 2014 study published in the journal *Neuron* found that curiosity puts the brain in a state that allows it to learn and retain information better (Gruber et al., 2014). When participants were curious about the answer to a question, they were better at learning that information. But remarkably, they were also better at learning unrelated information that was presented during their curious state.

This has profound implications for professional development: curiosity doesn't just help you learn what you're curious about—it enhances your capacity to learn everything.

WHY CURIOSITY IS A COMPETITIVE ADVANTAGE IN THE AI ERA

In 1986, psychologists Giyoo Hatano and Kayoko Inagaki introduced the concept of adaptive expertise—the ability to apply knowledge flexibly to new situations (Hatano & Inagaki, 1986). Their research distinguished between two types of experts:

- **Routine Experts**, who excel at executing well-established procedures with precision but struggle when confronted with novel challenges.
- **Adaptive Experts**, who not only master standard practices but also understand the underlying principles deeply, allowing them to modify their approach when circumstances change.

AI has mastered routine expertise. It can execute known procedures flawlessly, apply established patterns, and optimize within defined parameters. What it cannot do is question those parameters, challenge its own outputs, or imagine possibilities outside its training data.

This presents an opportunity hidden in plain sight: while everyone else is racing to become better at tasks AI already does well, the real advantage lies in cultivating the uniquely human capability to explore the unknown.

Reed Hastings didn't set out to kill Blockbuster—he asked a simple question: 'What if movies didn't have late fees?' That question led to Netflix. In every industry, the most valuable professionals aren't just executing tasks—they're asking the questions that create new possibilities.

AI's BUILT-IN LIMITATIONS: THE HUMAN OPPORTUNITY

1. AI answers but doesn't ask. AI systems can generate responses to queries, but they don't independently identify which questions matter. They don't wake up wondering *"What if?"* or feel compelled to challenge foundational assumptions. The strategic framing of questions remains a uniquely human domain (Marcus & Davis, 2019).

2. AI replicates patterns but doesn't reimagine them. AI excels at recognizing and extending existing structures, but it doesn't spontaneously question why those patterns exist or imagine radically different alternatives. It operates within the boundaries of its training data, unable to make the creative leaps that lead to paradigm shifts (Mitchell, 2019).

3. AI optimizes but doesn't discover purpose. AI can tell you the most efficient way to reach a goal, but it can't determine which goals are worth pursuing. It lacks the human ability to assign meaning, interpret ethical considerations, or align decisions with broader purpose (Tegmark, 2017).

Each of these AI limitations represents an opportunity for the curious human:

- While AI commoditizes answers, curious humans monopolize questions.
- While AI optimizes existing processes, curious humans reimagine them.

- While AI executes tasks, curious humans discover purpose.

This isn't about competing with AI—it's about transcending what AI does well to focus on what only humans can do:

- Explore with intention
- Question with purpose
- Connect disparate domains to create breakthrough insights

In a world where AI automates predictability, curiosity becomes the ultimate differentiator.

THE CURIOSITY ADVANTAGE FRAMEWORK: TURNING EXPLORATION INTO IMPACT

The difference between AI victims and AI architects comes down to one thing: who's asking the questions. Victims wait for questions to be assigned; architects generate their own. Victims execute within parameters; architects redefine the parameters. Victims use AI to do their existing job faster; architects use AI to create jobs that didn't exist before.

AI can generate variations on existing ideas. The AI Architect uses curiosity to discover the ideas AI can't generate at all.

The Compound Knowledge Advantage

There's another reason curiosity creates competitive advantage: it generates compound knowledge—insights that multiply rather than merely accumulate.

Linear knowledge grows by addition: 1 + 1 + 1 = 3. You learn a fact, then another fact, then another fact. This is what traditional education emphasizes, and it's what AI can do extremely well.

Compound knowledge grows by multiplication: 2 × 2 × 2 = 8. You learn a concept in one domain, connect it to a concept in another domain, and create an entirely new insight that didn't exist in either original domain.

Curious professionals build compound knowledge by constantly asking: "How does this connect to that?" They create value by linking previously unrelated dots—connecting psychology to marketing, biology to organizational design, physics to supply chain management.

This compound growth explains why interdisciplinary curiosity creates exponential advantages over linear expertise. Research on scientific breakthroughs shows that transformative innovations often come from cross-domain connections rather than deepening existing knowledge (Uzzi et al., 2013). The most valuable insights emerge not from mastering a single domain but from connecting multiple domains in novel ways.

Creating Advantage Through Curiosity Arbitrage

There's a specific form of advantage that curious professionals create in the AI era: curiosity arbitrage.

In finance, arbitrage refers to exploiting price differences between markets. In professional terms, curiosity arbitrage means exploiting knowledge gaps between domains that haven't yet been connected.

Here's how it works: When a field becomes heavily optimized and competitive, AI can quickly master the established patterns and commoditize the standard approaches. The curious professional doesn't try to compete in this crowded space. Instead, they look for adjacent fields where valuable insights haven't yet been applied.

For example:

- While everyone in digital marketing was optimizing the same metrics using the same AI tools, curious marketers were studying evolutionary psychology to understand deeper patterns of human attention.
- While financial analysts were using AI to optimize traditional market models, curious analysts were incorporating climate science into risk assessments.
- While designers were using AI to generate more visual options, curious designers were studying neuroscience to understand how design affects decision-making at a biological level.

The pattern is consistent: curiosity creates professional advantage by exploring gaps between domains before AI and competitors catch up.

THE CURIOUS MINDSET: THE 4 MENTAL SHIFTS THAT SEPARATE LEADERS FROM FOLLOWERS

Becoming an AI Architect rather than an AI victim requires more than just using the tools—it requires fundamentally shifting how you think. The curious mindset consists of four critical mental shifts that separate those who will thrive from those who will merely survive:

1. From Certainty to Exploration

The obsolete mindset: Expertise means having answers.

The curious mindset: Expertise means asking better questions.

The old professional paradigm rewarded certainty. Experts were people with answers. Leaders were decisive. Doubt was weakness.

That model is dead.

In an AI world, pretending to know everything is pointless when machines can access all human knowledge instantly. The new advantage lies in comfort with uncertainty—the willingness to say "I don't know, let's find out" instead of faking confidence.

The best leaders don't seek answers—they seek better questions. They create environments where not knowing is the beginning of discovery, not a sign of incompetence. They understand that in a world of constant change, certainty is often an illusion—and exploration is the only real security.

Research from Harvard Business School demonstrates that leaders who embrace intellectual humility and curiosity create teams that are more innovative, more agile, and better at problem-solving than teams led by leaders who emphasize certainty and confidence (Edmondson, 2018).

This isn't about being indecisive—it's about recognizing that exploration before decision-making produces better outcomes than premature certainty.

2. From Execution to Experimentation

The obsolete mindset: Master the standard operating procedure.

The curious mindset: Design experiments to create new procedures.

The industrial era trained us to execute—to follow procedures, meet metrics, and optimize processes. The AI era demands we experiment—to test hypotheses, embrace small failures, and learn continuously.

Curiosity isn't about passively consuming knowledge—it's about actively testing ideas. It's the difference between reading about swimming and jumping into the water.

While your colleagues are perfecting execution skills that AI will eventually automate, focus on building experimentation skills that AI cannot replicate: forming hypotheses about market trends, customer needs, or process improvements; designing small tests to validate your thinking; and rapidly incorporating what you learn into the next experiment.

Organizations that embrace experimentation consistently outperform those that prioritize flawless execution. A landmark study by McKinsey found that companies with strong experimentation cultures—marked by rapid test-and-learn approaches—achieved 2-3 times higher financial returns than industry peers over five to seven years (McKinsey & Company, 2018).

The ability to design and learn from experiments—both successful and failed—provides a sustainable advantage that AI cannot replicate.

3. From Problem Solving to Problem Finding

The obsolete mindset: Deliver solutions to defined problems. The curious mindset: Discover problems worth solving. Anyone can solve a clearly defined problem. The real value comes from finding problems worth solving—identifying opportunities, challenges, and questions that others haven't noticed.

Curious minds don't just solve problems—they discover new ones. They ask: What needs aren't being met? What assumptions are we making? What contradictions exist in our current approach? This isn't just creativity—it's strategic curiosity that reveals new market opportunities, uncovers hidden inefficiencies, and anticipates emerging challenges before they become crises.

Research published in the *MIT Sloan Management Review* shows that firms excelling at problem framing—identifying unaddressed needs—outperform problem-solving peers by 35% in innovation returns, a gap widened by AI's prowess at execution (Liedtka & Kaplan, 2023).

As AI systems become increasingly powerful problem-solvers, the human advantage will shift decisively toward problem finding—identifying which problems are worth solving in the first place.

4. From Depth to Breadth

The obsolete mindset: Go deep in one domain.

The curious mindset: Connect insights across many domains.

The specialist who goes deep into a single domain is increasingly vulnerable to AI disruption. The polymath who connects insights across multiple domains creates value AI cannot match. Curiosity compels us to explore beyond our specialties—to find connections between seemingly unrelated fields.

The financial analyst who understands climate science. The programmer who studies psychology. The designer who explores materials science. These cross-domain explorers develop unique perspectives that generate breakthrough insights.

This doesn't mean expertise is worthless—it means that singular expertise is insufficient. The new model is T-shaped knowledge: deep expertise in one area combined with broad awareness across many domains.

A groundbreaking study in *Science* analyzed 17.9 million scientific papers and found that the most impactful research paired deep conventional knowledge with novel insights from atypical fields, making these papers up to 3.7 times more likely to be among the most cited (Uzzi et al., 2013).

The pattern is clear: as AI gets better at optimizing within established domains, the greatest human advantage comes from exploring across domains.

THE CURIOSITY CHALLENGE: YOUR 30-DAY ACTION PLAN

Knowledge is only valuable when it changes behavior. So let's make curiosity actionable with The Curiosity Challenge—a structured 30-day program to develop curiosity as a professional practice rather than an occasional impulse.

Week 1: Curiosity Triggers

Day 1-2: Question Rotation

Start your workday by writing down three questions you don't have answers to—one about your industry, one about a customer/colleague, and one about an adjacent field. Keep these visible as you work.

Day 3-4: Assumption Auditing

List the five core assumptions underlying *your* work or industry. Challenge one assumption by asking: "What if this wasn't true anymore?" Explore the implications for ten minutes.

Day 5: Reverse Mentoring

Seek guidance from someone with less experience but a different perspective—a junior colleague, someone from another generation, or a professional from an entirely different field. Ask them what questions they have about your work that you've stopped asking.

Day 6-7: Pattern Interruption

Deliberately break your routine in ways that expose you to new stimuli. Take a different route to work, rearrange your workspace, or consume media from sources you wouldn't normally explore.

Week 2: Cross-Domain Exploration

Day 8-9: Adjacent Industry Study

Select an industry completely different from yours. Study how they approach a challenge similar to one in your field. Extract three principles that could transform your approach.

Day 10-11: Curiosity Partnership

Form a mutual curiosity alliance with someone in an adjacent field. Meet to share questions, insights, and resources. Establish a regular cadence for these exchanges.

Day 12-14: Deep Dive Session

Schedule a two-hour block of uninterrupted exploration on a topic you're curious about. Not for immediate work application, but for the sake of learning itself. Document key insights.

Week 3: Application & Experimentation

Day 15-16: Prototype Thinking

Convert one insight from your exploration into a small, testable application. Create a minimal viable prototype to test in your real work environment.

Day 17-18: Curiosity Loops

For every problem you solve this week, identify three new questions it raises. Document these questions and select one to explore further.

Day 19-21: Cross-Pollination Meeting

Organize a meeting with colleagues where everyone must bring one insight from outside the industry that might apply to a current challenge. Discuss potential applications.

Week 4: Integration & Amplification

Day 22-23: Value Translation

Practice explaining the value of your curious explorations to stakeholders in their language. Create a 2-minute pitch about how your explorations could create tangible business value.

Day 24-25: Insight Cataloging

Create a searchable repository of questions, insights, and connections that have emerged from your curiosity practice. Review it to identify patterns and potential value.

Day 26-28: Curiosity Metrics

Develop personal metrics to track your curiosity practice. Examples: Number of cross-domain insights generated monthly, experiments initiated, or assumption challenges that led to new approaches.

Day 29-30: Reflection & Commitment

Review your 30-day journey. What worked? What didn't? Which practices created the most value? Commit to continuing the 3-5 practices that were most impactful for you.

This 30-day challenge isn't just about building a habit—it's about rewiring how you approach your professional life. Even implementing a handful of these practices will progressively build your capacity for strategic curiosity.

THE CURIOUS PROFESSIONAL IN ACTION: CASE STUDIES

Let's see how the curious mindset translates into concrete professional advantage through three brief case studies:

Case Study 1: The Financial Analyst Who Reimagined Risk

Thomas, our financial analyst from earlier, applied the curious mindset to transform his approach to risk assessment. While his colleagues used AI to run standard risk models faster, Thomas got curious about what those models missed.

He began questioning the assumption that historical financial data was sufficient for future risk prediction. This led him to explore climate science research, where he discovered emerging methodologies for quantifying climate uncertainty.

By connecting financial modeling with climate science—two domains that rarely intersected—Thomas created a new approach to risk assessment that incorporated previously ignored variables. He didn't just improve existing risk models; he reimagined what risk modeling could be.

The result wasn't just better analysis—it was an entirely new service offering that positioned him as a pioneer rather than just another analyst. When his firm's traditional analysis roles were increasingly automated, Thomas was promoted to lead a new Climate Financial Risk division.

Case Study 2: The Marketer Who Discovered Precision Personalization

Maria was a mid-level marketing manager at a consumer packaged goods company when AI image generation and copywriting tools began disrupting her field. Instead of panic or resistance, she got curious about a fundamental assumption in her industry.

During an assumption audit, Maria challenged a core premise: "What if brand consistency isn't actually what consumers want in the age of personalization?" This led to her question: "How are other industries handling the tension between consistency and personalization?"

Maria conducted cross-industry research, particularly studying how precision medicine was transforming healthcare from standardized protocols to personalized treatment. She attended a healthcare innovation conference, built relationships with medical professionals, and studied outcomes in personalized care models.

This cross-domain curiosity led Maria to develop "Precision Marketing"—an approach that maintained core brand elements while using AI to dynamically personalize secondary elements based on individual consumer data.

The results were remarkable: the precision marketing approach drove a 34% increase in conversion compared to the company's traditional approach. What began as a curious exploration of an adjacent industry became a competitive advantage that transformed her company's entire marketing strategy.

Maria's career trajectory changed as well. When AI disrupted traditional marketing roles, she wasn't replaced—she was promoted to lead the new Precision Marketing division.

Case Study 3: The HR Director Who Revolutionized Recruitment

James was an HR director struggling with a persistent problem: despite implementing sophisticated applicant tracking systems and AI-powered resume screening, his company still suffered from high turnover and mediocre performance among new hires.

Instead of trying to optimize the existing recruitment process, James got curious about a different question: "What if we're measuring the wrong things in candidates?"

This question led him to explore research in fields seemingly unrelated to HR: complexity science, team cognition, and even improvisation theater. He discovered that while his company had been screening for credentials and experience (things AI could easily assess), they were missing crucial indicators of adaptability and collaborative intelligence.

By connecting insights from these diverse fields, James developed a completely new approach to talent assessment that focused on measuring a candidate's "adaptive potential" rather than just their history and credentials.

This curious exploration led to a revolutionary recruitment process that combined AI-driven assessments with uniquely human evaluation of qualities AI couldn't detect. The results were transformative: within 18 months, turnover decreased by 36% while new hire performance ratings increased by 28%.

When his company later implemented more advanced AI-driven HR systems, James wasn't made obsolete—he became more valuable as the architect of human-AI collaborative recruitment strategies.

ORGANIZATIONAL CURIOSITY: BEYOND INDIVIDUAL PRACTICE

While The Curiosity Challenge works at the individual level, the truly curious professional also considers how to foster curiosity across teams and organizations.

Consider these approaches for scaling curiosity beyond your individual practice:

Curiosity Reviews

In addition to performance reviews, implement quarterly curiosity reviews where team members share the questions they're exploring, what they're learning from adjacent fields, and how those insights might create value.

Exploration Time

Formalize time for curiosity, following the 20% model pioneered by innovation-leading companies. Dedicated exploration time isn't a luxury—it's a strategic investment in future value creation.

Question-Driven Meetings

Transform stale meetings with a simple protocol: every participant must bring one genuine question related to the topic at hand. Start meetings with these questions rather than updates or presentations.

Cross-Functional Curiosity Teams

Create temporary teams drawn from different departments with the explicit purpose of exploring a question or challenge from multiple perspectives. These teams aren't focused on immediate execution but on generating novel insights.

Learning Exchanges

Implement structured programs where employees spend time in different departments or even different companies to gain fresh perspectives and bring back questions and insights.

Organizations that implement these practices don't just become more innovative—they become more resistant to AI disruption. Research from MIT's Center for Collective Intelligence demonstrates that organizations with structured cross-functional exploration significantly outperform peers in adaptation to technological disruption (Malone & Bernstein, 2022).

When curiosity becomes a cultural norm rather than an individual trait, the entire organization becomes capable of evolving ahead of technological change rather than reacting to it.

THE CURIOUS WILL WIN: YOUR CALL TO ACTION

The AI revolution isn't rendering humans obsolete—it's creating the greatest opportunity in history for curious minds to thrive.

While machines master the known, humans will master the unknown. While AI optimizes existing patterns, the curious will imagine new ones. While AI answers yesterday's questions, the curious will discover tomorrow's.

In a world where knowledge is increasingly commoditized, the ability to explore will always beat the ability to execute. Curiosity isn't just how you stay relevant—it's how you stay irreplaceable.

Here's your call to action:

1. **Start your Curiosity Challenge today.** Choose three practices from the 30-day plan and implement them this week. Don't wait for permission or the perfect moment.
2. **Identify your exploration domain.** What adjacent field, if connected to your expertise, could create breakthrough value? Commit to exploring it systematically.

3. **Build your curiosity alliance.** Find at least one colleague or connection who will join you in regular curiosity exchanges. Accountability accelerates growth.
4. **Measure your curiosity impact.** Track how your explorations translate into tangible value—new approaches, better solutions, or emerging opportunities.

Learning new skills is meaningless if you don't escape the gravitational pull of outdated career structures. AI isn't just changing what you need to know—it's dismantling the entire way expertise is recognized, rewarded, and monetized. The professionals who win won't just collect skills; they'll position themselves in ways that make them irreplaceable. Let's break the anti-ladder wide open.

But curiosity alone isn't enough. The professionals who thrive in the AI era aren't just those who ask better questions—they're the ones who strategically build AI-proof skill stacks. Let's break down how to combine your capabilities in a way that creates career resilience no algorithm can replicate.

Later in this book, curiosity reappears as one of the seven superpowers AI can't replicate. But make no mistake: it's not just a skill—it's the ignition switch. Without curiosity, the rest of your AI-proof capabilities never fully activate. This chapter planted the seed. You'll see how it grows into something even more powerful in Part 3

THREE THINGS FOR THIS WEEK

Reality Check: It's Not the Smartest Who Win—It's the Fastest Adapters

1. Have one conversation this week about **how AI is impacting your industry**—compare insights.
2. Challenge yourself to **learn one skill outside your domain** that increases adaptability.
3. If your job were disrupted today, what would be your **next move**? Outline a rough backup plan.

07

Forget Career Paths
Stack AI-Proof Skills Instead

Ladders are dead. The future belongs to skill stackers.

The future belongs to the curious. Not the smartest. Not the hardest working. The curious. Why? Because when AI can know everything, the only people who matter are the ones who explore beyond it. The ones who ask the questions no machine would think to ask.

AI isn't just automating tasks—it's rewriting the rules of career success. Roles that were once stable for decades now shift in years, sometimes months. The professionals who thrive aren't those with the most knowledge, but those who integrate knowledge in ways AI can't replicate. That's where skill stacking comes in.

If linear career paths are dead, what replaces them? The answer isn't a new map—it's a new mindset: skill stacking, or what we call The Adaptive Stack.

The Adaptive Stack: Thriving in an AI-Powered World

AI is redefining the landscape of work, making traditional specialization increasingly vulnerable to automation. To stay ahead, professionals need more than deep expertise in a single field—they need a system for adaptability. That system is The Adaptive Stack.

The Adaptive Stack is the deliberate combination of complementary skills, mindsets, and knowledge domains that create exponential value at their intersection. Unlike traditional career paths that emphasize linear specialization, The Adaptive Stack is dynamic, evolving, and designed to thrive amid technological disruption. Those who master it become indispensable architects of innovation, rather than executors of predefined roles.

At its core, The Adaptive Stack is built on three foundational layers:

Foundation Tier: Transferable Meta-Skills These are the skills that never go obsolete, regardless of industry or technology shifts:

- Systems thinking
- Critical reasoning
- Learning agility
- Communication across boundaries
- Adaptability

Specialization Tier: Domain-Specific Expertise The strongest Adaptive Stacks consist of three or more fields that intersect to create unique, AI-resistant value:

- Technical domains (e.g., data science, UX design)
- Process domains (e.g., agile methodologies, design thinking)
- Industry domains (e.g., healthcare, fintech)

Application Tier: Synthesis and Deployment This is where knowledge turns into action—the ability to apply expertise to real-world challenges and generate meaningful impact:

- Translating concepts into outcomes
- Adapting approaches for different contexts
- Measuring and communicating impact

By layering these tiers effectively, professionals future-proof themselves, staying ahead of AI-driven disruption while continuously expanding their combinatorial advantage.

Would your current skill set stand the test of AI? If your expertise disappeared tomorrow, could you pivot without starting from scratch?

The Adaptive Stack ensures the answer is always YES.

THE END OF THE SOLO PERFORMANCE

Picture your career as it once was: a single instrument playing up a predictable scale. For years, success meant mastering that instrument—marketing, coding, analysis—and climbing a well-defined ladder. Then AI arrived on stage, hitting those same notes faster, louder, and often better.

That solo performance? It's ending. But you're not being replaced—you're being promoted to conductor.

In this new reality, your future isn't in outplaying AI at what it does best. It's in orchestrating a combination of skills that creates value no algorithm can replicate alone.

Maya discovered this during her annual review.

"What's your career path?" the senior executive asked Maya.

Five years earlier, she would have had a ready response—a linear progression from her current marketing role to senior manager, then director, and eventually CMO. It was the expected answer, the one that showed ambition and clarity.

Instead, she said something unexpected: "I'm not focused on a path. I'm building a skill stack."

The executive looked confused. "A what?"

"Rather than climbing a predefined ladder," Maya explained, "I'm developing a unique combination of capabilities that create value no matter how our industry evolves. Right now I'm combining marketing strategy, data analytics, and behavioral psychology in ways that help us connect with customers more effectively. As AI transforms our field, this combinatorial approach creates more resilience than any linear career path."

The executive was silent for a moment. Then he smiled. "That's the smartest answer I've heard in twenty years of doing these reviews."

Six months later, when the company restructured its marketing department around AI-powered systems, many traditional roles were eliminated. Maya wasn't just retained—she was tapped to lead a new customer intelligence unit that hadn't existed before.

Her unique skill combination was suddenly more valuable than any traditional marketing title.

"People who defined themselves by their position on the old path were vulnerable," Maya reflected later. "People who had built distinctive skill stacks became indispensable in the new reality."

Maya's experience illustrates the central argument of this chapter: In the AI era, success isn't about climbing predetermined career ladders. It's about building unique skill combinations that create value no algorithm can replicate.

This isn't the first time technology has upended careers. The Industrial Revolution displaced artisans with machines. Software automation eliminated entire categories of administrative work. Outsourcing redefined white-collar stability. But AI is different—it's not just moving jobs, it's dismantling expertise itself. While previous technological revolutions mainly affected physical production or routine information processing, AI strikes at the heart of knowledge work—the very domain where most professionals have sought refuge.

The traditional career path—with its linear progression through predefined roles—is a relic of a more stable economic era. In a world of accelerating change and AI-driven disruption, this outdated model creates vulnerability rather than security.

The alternative isn't chaos or random skill acquisition. It's a strategic approach to professional development that leverages The Adaptive Stack—a system designed specifically to thrive in an AI-powered world.

Why Single-Domain Experts Are Endangered

By 2030, pure specialists in most fields may find themselves either unemployed or subordinate to algorithms. For generations, the path to professional success was straightforward: select a domain, specialize deeply, and advance through increasingly senior roles based on accumulated expertise.

This model thrived when: (1) knowledge evolved slowly within well-defined boundaries, (2) human specialists were the primary source of domain-specific expertise, and (3) organizations maintained stability sufficient for predictable career progression.

Artificial intelligence has disrupted all three conditions. Today: (1) knowledge boundaries shift and blur at an unprecedented pace, (2) AI systems can access, synthesize, and apply domain knowledge instantaneously, and (3) organizations continuously restructure around technological advancements.

Evidence of this transformation is mounting. AI systems now rival or surpass human performance across a growing array of specialized fields: AI-driven diagnostic tools outperform radiologists in detecting breast cancer (McKinney et al., 2020); legal AI platforms review contracts with 94% accuracy in seconds, surpassing seasoned attorneys (Surden, 2019); AI-powered financial systems process vast datasets in real time, outpacing human analysts (Huang et al., 2022).

A 2023 analysis confirms this shift is accelerating, with AI dominating structured domains and reshaping work across industries (Brynjolfsson et al., 2023).

Carlos experienced this disruption firsthand. After fifteen years building his expertise as a financial analyst, he'd earned advanced certifications, specialized in a lucrative niche, and was recognized as one of the top talents in his field. Then AI-powered analysis tools began transforming his industry almost overnight.

"I watched competitors adopt AI systems that could perform analyses in minutes that would take me days," Carlos recalled. "At first, I dismissed them—there's no way an algorithm could match my specialized knowledge. Then I saw the outputs. The AI wasn't just faster—in many cases, it was more thorough."

This realization sparked a critical period of reflection. "I'd been defining my value through a single domain of expertise," Carlos explained. "I suddenly realized this was incredibly risky. If my core skill was being automated, what exactly was I offering?"

Rather than doubling down on his existing specialty or starting over in a new field, Carlos took a different approach: he began deliberately combining his financial expertise with other domains.

"I started exploring behavioral economics to understand the human factors behind financial decisions," he explained. "Then I added sustainability expertise to analyze environmental impacts on long-term financial performance. Finally, I developed skills in communication design to translate complex analyses into actionable insights for decision-makers."

This strategic combination created something AI couldn't replicate: the ability to synthesize insights across domains in contextually relevant ways. "My value isn't just financial analysis anymore—it's connecting financial patterns to human behavior and sustainability trends, then communicating those connections in ways that drive better decisions."

When his firm restructured around AI systems, Carlos wasn't replaced—he was promoted to lead a new division focused on integrated decision support that combined AI-powered analysis with human synthesis across domains.

"The specialists who clung to their single domain were the most vulnerable," Carlos observed. "They were competing directly with AI on its terms. Those of us who built unique combinations created value no algorithm could match."

The secret to Carlos's success wasn't abandoning his primary expertise—it was strategically combining it with complementary domains that created unique value at their intersection.

WHY COMBINATIONS CREATE MORE VALUE THAN SPECIALIZATION

For decades, professional development followed a simple formula: Specialize in a valuable domain, become increasingly expert within that domain, and advance through predefined roles based on that expertise.

This approach worked because:

- Domains evolved relatively slowly
- Specialization created scarcity value
- Organizational structures remained fairly stable

The AI era has disrupted all three conditions. Domains transform rapidly, specialized knowledge becomes accessible to anyone (including algorithms), and organizations continuously restructure around technological capabilities.

In this new environment, pure specialization creates vulnerability rather than security. The deeper your expertise in a single domain, the more exposed you become to disruption when that domain transforms—which it inevitably will.

In an AI-powered world, the old career playbook no longer works. The solution? Building unique combinations that AI can't replicate.

Consider these high-value skill combinations that create resilience amid AI disruption:

- Financial analysis + sustainability metrics + regulatory expertise
- Product design + behavioral psychology + data visualization
- Software development + healthcare operations + patient experience
- Supply chain logistics + predictive analytics + geopolitical risk assessment

The pattern is clear: Three or more skill domains combined create exponentially more value than deeper expertise in just one. And crucially, these combinations are less vulnerable to AI disruption because they create value through synthesis and novel application rather than routine expertise.

This doesn't render all human expertise obsolete; rather, it redefines what expertise is valuable. In an economy where algorithms can instantly master domain-specific knowledge, single-domain expertise offers diminishing returns and heightened risk of obsolescence.

The emerging premium lies in integration expertise—the capacity to synthesize knowledge from multiple domains into novel, high-value solutions. This is distinct from shallow generalism. It requires substantial depth in several fields and the ability to generate unique insights through their intersection.

Want to test your career resilience? Ask yourself:

- Do I have at least three distinct skill domains that I can combine?
- Can I apply my expertise across multiple industries?
- If my job disappeared tomorrow, could I pivot into something new without starting from scratch?
- Has my skill combination ever helped me solve a problem that stumped specialists?

If you answered "no" to any of these questions, your skill stack may not be AI-proof yet. Let's fix that.

BUILDING YOUR AI-RESISTANT ADAPTIVE STACK

Not all skill combinations create equal value or resilience. A truly effective Adaptive Stack has specific architectural characteristics:

1. Complementary Rather Than Random

Elena, a graphic designer, watched with growing anxiety as AI image generators began producing work that rivaled what took her hours to create. "I spent a decade perfecting my craft, and suddenly it felt like the foundation was crumbling," she recalled.

Instead of either abandoning her design expertise or ignoring AI's advance, Elena deliberately constructed a skill stack with strategic combinations.

"I realized that adding random capabilities wouldn't help," she explained. "I needed domains that would multiply each other's value." Elena combined her design expertise with user psychology and data visualization—three domains that created exponentially more value together than separately.

2. Technical and Human-Centered Balance

"While developing data visualization techniques, I also deepened my understanding of emotional design and storytelling," Elena noted. "The technical skills helped me create precise information displays, while the human-centered skills ensured they connected emotionally with audiences."

3. Depth and Breadth

Elena developed both depth and breadth. "I maintained deep expertise in design fundamentals while developing working knowledge across adjacent domains," she explained. "This combination let me lead projects that required specialized design knowledge while understanding enough about other domains to integrate them effectively."

4. Established and Emerging Domains

"I kept refining my core design principles—those fundamentals haven't changed in centuries," Elena said. "But I also developed capabilities in emerging fields like interactive data systems and augmented reality prototyping." This combination created both current value and future options.

5. Explicit and Tacit Knowledge

"Beyond the technical skills that can be taught, I developed intuitive capabilities that come only through experience," she reflected. "My ability to sense when a design 'feels right' or to anticipate how users will emotionally respond to a visualization—these tacit capabilities are nearly impossible for AI to replicate."

The result was an Adaptive Stack with remarkable resilience. When her company restructured around AI design tools, Elena wasn't replaced—she became the leader of a new "human-AI design collaboration" team that leveraged both machine efficiency and human creativity.

"What saved me wasn't clinging to traditional design or simply learning to use AI tools," Elena concluded. "It was deliberately building a unique Adaptive Stack that created value through synthesis across domains."

Elena's experience illustrates how a deliberate Adaptive Stack creates resilience that random skill collection cannot match. The specific combination matters far more than the individual components.

THE THREE-TIER FRAMEWORK IN ACTION

This three-tier framework of The Adaptive Stack isn't just theoretical—it transforms careers in practical ways, as James discovered during AI's rapid transformation of the software development field.

FOUNDATION TIER: TRANSFERABLE META-SKILLS

James started with these indispensable capabilities that retain value regardless of technological change:

- Systems thinking
- Critical reasoning
- Learning agility
- Communication across boundaries

- Adaptability

SPECIALIZATION TIER: DOMAIN-SPECIFIC EXPERTISE

For this tier, James deliberately selected three complementary domains that would create unique value at their intersection:

- Technical domains (e.g., data science, UX design)
- Process domains (e.g., agile methodologies, design thinking)
- Industry domains (e.g., healthcare, fintech)

APPLICATION TIER: SYNTHESIS AND DEPLOYMENT

Finally, James developed these capabilities for turning knowledge into tangible value:

- Translating concepts into outcomes
- Adapting approaches for different contexts
- Measuring and communicating impact

James was witnessing AI's impact on his field of software development when he decided to reimagine his professional capabilities using the three-tier framework.

"I realized my coding skills alone—no matter how advanced— would increasingly compete with AI systems," James explained. "I needed a more structured approach to developing my capabilities."

He began with the Foundation Tier. "Instead of just focusing on specific programming languages, I deliberately developed systems thinking across multiple domains," James recalled. "I practiced critical reasoning about architecture decisions, improved my

175

communication across technical boundaries, and cultivated learning agility that let me master new technologies quickly."

These meta-skills provided the base for his skill stack—capabilities that would remain valuable regardless of which specific technologies emerged or faded. "When a new framework replaces an old one, my systems thinking helps me see patterns that others miss," James noted. "While others struggle with the syntax, I grasp the underlying architecture quickly."

Next, James built his Specialization Tier. "I developed deep knowledge in three complementary domains: distributed systems architecture, machine learning operations, and cybersecurity," he explained. "Each domain was valuable independently, but the real power came from their intersection."

James chose these specializations strategically. They complemented each other in non-obvious ways, included areas less vulnerable to near-term automation, and represented different types of knowledge—technical fundamentals, operational practices, and security principles.

"The magic happens at the intersections," James observed. "Understanding how machine learning models behave in distributed systems, or how security concerns affect architectural decisions—these combinations create insights that specialists in any single domain often miss."

Finally, James developed his Application Tier. "Having knowledge isn't enough—you need to apply it to create tangible value," he explained. "I developed specific capabilities for translating technical concepts into business outcomes, adapting approaches for different stakeholders, and measuring impact in ways decision-makers could understand."

This application tier ensured James's knowledge didn't remain theoretical but created practical value in specific contexts. "I've seen brilliant technologists struggle because they can't translate their expertise into solutions that address real business problems," James noted. "The application tier bridges that gap."

The three-tier approach transformed James's career trajectory. When AI coding assistants began handling routine development tasks, he wasn't threatened—he was positioned to lead AI-human collaboration initiatives that required precisely his unique combination of capabilities.

"What protected me wasn't just knowledge accumulation—it was deliberate knowledge architecture," James concluded. "The three-tier framework gave me structure that random skill collection never could."

THE HUMAN+AI SKILL STACK: BECOMING AUGMENTED, NOT AUTOMATED

The most resilient skill stacks don't compete with AI; they complement it. This requires developing a "Human+AI" skill configuration—capabilities that work with rather than against advancing technology.

The key principle is focusing on augmentation rather than automation. While automation replaces human activity, augmentation combines human and machine capabilities to create outcomes neither could achieve alone.

Research confirms that workers who master this augmentation approach—blending AI fluency, human judgment, boundary-spanning, procedural know-how, and orchestration—consistently outperform those reliant on automation alone (Jarrahi et al., 2022).

Priya, a research analyst at a consulting firm, watched with growing concern as AI systems began generating reports that seemed comparable to her own work. Instead of either ignoring AI or surrendering to it, she deliberately developed a Human+AI skill configuration that transformed her relationship with technology.

"The turning point came when I stopped seeing AI as competition and started seeing it as a collaborator," Priya explained. "This required developing five specific capabilities that changed everything."

1. AI Fluency Without AI Dependency

"I learned enough about different AI systems to leverage them effectively without becoming dependent on any specific platform," Priya explained. "When one tool was replaced by something better, I could quickly adapt because I understood the underlying principles, not just button sequences."

This fluency allowed Priya to evaluate and adopt new tools rapidly, maintaining technological currency without becoming locked into obsolete systems.

2. Uniquely Human Capabilities

"I stopped trying to compete with AI on information processing and pattern recognition," Priya recalled. "Instead, I doubled down on contextual judgment, creative synthesis, and ethical reasoning—areas where even advanced AI systems struggle."

This emphasis on distinctly human strengths meant Priya was enhancing rather than duplicating what AI could do, creating complementary value rather than competitive conflict.

3. Boundary-Spanning Expertise

"I developed the ability to translate between AI capabilities and organizational needs," she explained. "I could understand both what the technology could do and what people actually needed, then design bridges between them."

This connective expertise positioned Priya at crucial intersections where technical and human domains met—precisely where the greatest value was created.

4. Procedural Knowledge

"AI systems can access vast information about principles and concepts," Priya noted. "But knowing how to apply that knowledge in messy, real-world situations with conflicting priorities and ambiguous data—that's where humans still excel."

By focusing on know-how rather than just know-what, Priya developed capabilities that were inherently more resistant to automation than fact-based expertise.

5. AI Orchestration Capabilities

"I created workflows where AI handled data processing and pattern recognition while humans provided contextual understanding and ethical oversight," she explained. "These hybrid systems produced better outcomes than either humans or AI could achieve alone."

This orchestration expertise positioned Priya not as someone who either used AI or competed with it, but as someone who designed how humans and AI could work together optimally.

The results transformed Priya's career trajectory. "When my colleagues were worrying about being replaced, I was being promoted to lead our new 'Augmented Intelligence' practice," she recalled. "What protected me wasn't resisting AI but becoming someone who could maximize its value while adding the human elements it couldn't provide."

FROM PROFESSIONAL PROFILES TO ADAPTIVE STACKS

Sophia's career journey illustrates the evolution from specialization to an Adaptive Stack—and the increasing resilience this transition created.

"I started my career with deep expertise in marketing analytics," Sophia explained. "This served me well in a stable environment where that specialty was valued and rarely disrupted."

As AI began transforming marketing analytics, Sophia recognized the vulnerability of single-domain depth. "I realized I needed additional areas of specialized expertise," she recalled. "I deliberately developed depth in behavioral psychology to complement my analytics background."

"Having expertise in both analytics and psychology created unique value," she explained. "I could analyze customer data patterns and understand the psychological drivers behind them—a combination neither pure analysts nor psychologists typically possessed."

Over the next few years, she deliberately developed meaningful depth in three additional domains: emerging technology evaluation, cross-cultural communication, and change management.

"Each specialty creates value independently," she explained, "but the real power comes from their unique combination. I can analyze behavioral data across cultures, evaluate technology impacts on customer psychology, and design change processes that account for both analytical insights and human resistance."

This diverse Adaptive Stack created remarkable resilience. When any single domain faced disruption, Sophia's expertise combination maintained her value. "When AI started generating cross-cultural communication recommendations, I wasn't threatened because that was just one element of my stack," she noted. "I could integrate those AI outputs with my other domains to create synthesis no algorithm could match."

The progression to a diverse Adaptive Stack wasn't accidental. "This was a deliberate evolution based on recognizing increasing vulnerability," Sophia reflected. "Each stage created more resilience than the last by developing unique combinations that became progressively harder for either humans or AI to replicate."

BEYOND THE LADDER: THRIVING WITHOUT PREDETERMINED PATHS

The death of traditional career ladders creates both challenge and opportunity. The challenge is navigating without the clarity of predefined progression. The opportunity is creating greater value—and capturing more of that value yourself—than was possible in rigid organizational structures.

Thriving in this new landscape requires a fundamentally different approach to professional development and value creation.

From Role Occupation to Value Creation

The core mindset shift is moving from "occupying roles" to "creating value"—regardless of formal position or title.

This means:

- **Identifying Value Gaps**: Continually scanning for spaces where your unique skill combination could solve problems or create opportunities others don't see.
- **Creating Before You're Asked**: Developing solutions to these challenges whether or not they fall within your formal responsibilities.
- **Demonstrating Impact Tangibly**: Creating clear documentation of the value your unique skill combination delivers.
- **Capturing Appropriate Value**: Ensuring your compensation reflects the unique value you create rather than standardized role benchmarks.

This shift benefits both organizations and professionals. Organizations access more innovative solutions as people apply their unique skill combinations beyond rigid role boundaries. Professionals create and capture more value than would be possible within predetermined paths.

FROM CAREER PLANS TO ADAPTIVE OPTIONS

Rather than following linear career plans, Adaptive Stack professionals develop what strategists call "options portfolios"— multiple possible paths they can activate as conditions evolve.

Building a robust options portfolio involves five specific practices:

- **Maintain Professional Experiments**: Always have 2-3 small projects outside your primary work that explore new domains or applications.
- **Develop Minimal Viable Skills**: Build sufficient capability in emerging areas to evaluate their potential without major investment.
- **Create Diverse Network Connections**: Cultivate relationships across different industries and functions that provide visibility into varied opportunities.
- **Build Public Evidence**: Create tangible artifacts that demonstrate your capabilities beyond your current role or organization.
- **Maintain Financial Flexibility**: Structure your finances to enable strategic pivots when valuable opportunities emerge.

These practices create resilience not through rigid planning but through continuous option development. When disruption occurs—whether from technological change, economic shifts, or organizational transformation—you have multiple paths already partially developed rather than a single vulnerable trajectory.

YOUR ADAPTIVE STACK ACTION PLAN

Top 5 AI-Proof Career Strategies

1. **Redefine Your Value Through Outcomes, Not Tasks**
   - Core Principle: Tasks are increasingly automatable; outcomes require orchestration of multiple capabilities.
   - Key Action: Transform how you articulate your professional contribution from "I do X" to "I enable outcome Y."
   - Success Indicator: You can clearly explain your value without mentioning specific activities or tools.

2. **Build a Distinctive Skill Stack, Not a Deep Specialty**
   - Core Principle: Unique combinations of 3+ skills create exponentially more value than deeper expertise in one domain—this is the essence of the Adaptive Stack.
   - Key Action: Identify complementary capabilities that create multiplicative rather than additive value when combined, focusing on cross-domain synthesis.
   - Success Indicator: You can name at least three domains where you have meaningful depth that rarely appear together.

3. **Develop Adaptation Velocity as Your Meta-Capability**
   - Core Principle: The speed at which you can adjust to changing conditions is your ultimate competitive advantage.

o   Key Action: Systematically build learning agility, identity flexibility, comfort with ambiguity, and recovery resilience.

o   Success Indicator: You consistently incorporate new approaches faster than peers and rebound quickly from setbacks.

4. **Position Yourself as an AI Architect, Not a Task Executor**

o   Core Principle: The highest value comes from designing systems, not performing tasks within them.

o   Key Action: Shift focus from execution excellence to system design—creating frameworks that optimize outcomes.

o   Success Indicator: Your work increasingly involves designing approaches rather than just implementing them.

5. **Create a Perpetual Transformation System, Not Just Adaptation Plans**

o   Core Principle: Continuous reinvention requires systematic approaches, not just willingness to change.

o   Key Action: Build integrated practices for horizon scanning, learning acceleration, identity evolution, recovery systems, and opportunity architecture.

o   Success Indicator: You thrive through change rather than despite it, leveraging disruption as a catalyst for growth.

THE FUTURE BELONGS TO ADAPTIVE STACKERS

The traditional career model offered clarity but limited optionality. You knew exactly what the next rung looked like, but your path was constrained by organizational structures and industry conventions.

The Adaptive Stack approach offers the opposite tradeoff: less predefined clarity but dramatically expanded optionality. You may not know exactly what your role will be in five years, but you'll have developed capabilities that create value across a wide range of possible futures.

The difference between those who become obsolete and those who thrive isn't whether AI impacts their field—it's how they design their Adaptive Stack to stay relevant, indispensable, and ahead of the curve.

In a world of accelerating change—where entire industries transform and new ones emerge with unprecedented speed—optionality consistently outperforms predictability. The professionals who thrive aren't those who execute predetermined paths most efficiently; they're those who develop the most robust set of options for creating value amid uncertainty.

This isn't just a defensive strategy for surviving disruption. It's an offensive approach to creating extraordinary value—and extraordinary professional success—by developing combinations that solve problems in ways no one else can. Your Adaptive Stack becomes your competitive advantage in a world where narrow expertise alone is increasingly vulnerable.

This isn't about learning to use AI tools superficially—it's about developing the combinatorial intelligence to know which problems AI can solve, which require human intervention, and how to blend the two for maximum impact. In the AI era, your unique skill stack is your only true competitive advantage.

This is the choice you face: evolve into an AI-augmented skill stacker or become algorithmically irrelevant. The difference between being indispensable and invisible comes down to whether you build combinations that machines can't easily replicate.

Your curiosity is your power. But only if you use it. AI is already answering questions at scale. The only way to stay ahead is to ask the questions no one else is asking.

Will you?

THREE THINGS FOR **THIS** WEEK

Reality Check: Building Your Adaptive Stack—Starting Now

1. Experiment with AI for creativity or problem-solving, not just automation—note the results and identify where human-AI collaboration creates unique value.
2. Assess your professional network—are you connected to people with diverse domain expertise? If not, engage with one new cross-disciplinary group.
3. Define your Adaptive Stack strategy—identify three complementary domains where you can develop expertise that, when combined, creates something AI cannot easily replicate.

PART 3

YOUR AI-PROOF PLAYBOOK

Now it's time to execute. This section delivers the tactical strategies to not just survive, but dominate in an AI-driven world.

WHERE WE'RE HEADED

Understanding AI is one thing. Taking action is another. This section is about execution—how to put everything you've learned into practice. You'll get a clear, step-by-step strategy to future-proof your career, integrate AI into your work, and continuously evolve so you stay ahead no matter how fast things change. This isn't about keeping up; it's about leading the way. By the time you finish this section, you won't just be AI-aware—you'll be AI-proof.

08

THE UNIQUELY HUMAN EDGE
THE SUPERPOWERS THAT AI CAN'T REPLACE

AI is the ultimate force multiplier—but only for those who master their human superpowers.

"The real danger is not that machines will begin to think like humans, but that humans will begin to think like machines."
— Sydney J. Harris

To thrive in an AI-powered world, you don't need superhuman intelligence—you need superhuman adaptability.

The question isn't whether AI will disrupt your industry—it already has. The only question that matters is: Are you leveraging the human edge AI can't touch, or are you betting your future on tasks AI is already learning to do better, faster, and cheaper?

The world is not dividing between those who understand AI and those who don't. The true separation is emerging between those who recognize their uniquely human capabilities and those who compete with machines on machine terms. This distinction will define the next era of work, wealth, and influence.

In a landscape increasingly populated by intelligent systems, your competitive advantage isn't technical prowess—it's your humanity (Brynjolfsson & McAfee, 2022). This isn't about resisting technology but rather understanding the new dynamic: AI is the ultimate force multiplier—but only for those who master their human edge.

Recent advances in generative AI have commoditized tasks once considered untouchable by automation—creative writing, visual design, data analysis, and even basic programming (Eloundou et al., 2023).

What once took years to become obsolete now takes months. The half-life of professional skills has shrunk to about five years (World Economic Forum, 2023). Yet amid this acceleration, a counterintuitive truth emerges:

The more powerful AI becomes, the more valuable uniquely human attributes become. The very qualities that machines struggle to replicate—curiosity, empathy, judgment—are precisely what will differentiate the irreplaceable from the automatable (Davenport & Ronanki, 2018).

This chapter introduces what we call the "Seven Superpowers"—the human capabilities that AI cannot replicate and that will define success in an AI-saturated world. These aren't just nice-to-have skills; they are the new non-negotiables for anyone who wants to remain relevant, valued, and employed.

We explored curiosity earlier as the edge that drives reinvention. Here, it returns not just as a mindset—but as one of the core superpowers. It's what fuels your ability to ask better questions, spot unseen patterns, and navigate ambiguity when AI hits a wall. If you've been cultivating curiosity since Chapter 6, you've already started building the muscle.

THE SEVEN HUMAN SUPERPOWERS: YOUR COMPETITIVE ADVANTAGE IN AN AI WORLD

1. Curiosity — AI Knows, But Humans Wonder

Think about those moments—those innately human flashes—when you feel a tiny spark of an idea. A subtle tingle, almost a twitch. That's your curiosity waking up, the first signal of something new, something worth exploring. It's the beginning of creativity at work. And no matter how powerful AI becomes, it will never experience that moment. That impulse to question, to wonder, to push beyond the expected—that's uniquely human. The key is learning to trust it, sharpen it, and act on it before complacency takes over. AI has infinite answers. Humans ask better questions. AI retrieves. We explore.

The most AI-proof professionals are the ones who challenge assumptions and push boundaries AI can't see. Artificial intelligence excels at providing answers within defined parameters, yet it remains fundamentally reactive—unable to wonder or question its own assumptions (Marcus & Davis, 2023).

The most transformative breakthroughs in human history began not with answers but with questions that challenged conventional thinking, a trait common among innovators from Ada Lovelace to Steve Jobs (Isaacson, 2014).

Consider quantum physics pioneer Richard Feynman, who didn't simply memorize formulas but questioned fundamental assumptions about how the universe works. His famous quote captures this superpower perfectly: "I would rather have questions that can't be answered than answers that can't be questioned" (Gleick, 2011, p. 289). While large language models can generate plausible-sounding answers, they struggle to identify novel research questions, instead favoring incremental advances within existing frameworks, as a 2023 study found (Wu & Dredze, 2023).

Behavioral scientist Todd Kashdan's research frames curiosity as a "knowledge emotion" with clear neurobiological underpinnings. His studies show that highly curious individuals exhibit heightened activity in reward-anticipation brain regions, such as the ventral striatum, when encountering novel stimuli—a pattern that fuels exploration beyond the familiar (Kashdan, 2009; Kang et al., 2009). Kashdan argues that curiosity "bridges the known and the unknown," forging neural pathways that AI, despite its computational prowess, cannot replicate due to its lack of intrinsic human motivation.

When Marie Curie pursued her investigations into radioactivity—working in a shed with primitive equipment and facing institutional barriers as a woman scientist—she wasn't responding to external prompts or optimizing for known parameters. She was driven by what psychologists now identify as "epistemic curiosity"—the intrinsic desire to close gaps in understanding, even at personal cost. This form of curiosity, deeply connected to human identity and purpose, transcends the pattern-matching capabilities of even the most advanced machine learning systems.

To strengthen this superpower, Wade (2023) recommends:

1. Practice question storming instead of brainstorming—generate 20 questions about a problem before seeking solutions
2. Adopt a beginner's mind by regularly entering unfamiliar domains where your expertise doesn't apply
3. Establish intellectual friction by seeking out those who disagree with your core assumptions
4. Develop "productive discomfort" with uncertainty by deliberately exploring areas where answers remain elusive or incomplete

2. Empathy — AI Simulates, But Humans Connect

AI mimics emotion, but humans build trust. AI can analyze sentiment, but it fundamentally does not experience emotions (Marcus, 2023). This distinction creates what philosopher Thomas Nagel might call an "empathy gap"—AI can process the data of human experience without accessing the subjective reality of what it means to be human.

This limitation matters profoundly because trust—the foundation of effective leadership, healthcare, education, and countless other human interactions—requires genuine empathy. Research demonstrates that even when AI mimics empathetic language, humans detect the absence of authentic emotional connection, creating an "uncanny valley of trust" (Thomason & Williams, 2023).

Consider healthcare, where studies show that a physician's empathy correlates directly with patient outcomes—treatment adherence improves by 38%, and recovery times decrease by 25% (Kim et al., 2022).

Neuroscientist Marco Iacoboni's research demonstrates that empathy operates through mirror neuron pathways that AI—lacking biological embodiment—cannot replicate, grounding our understanding in lived emotional experience (Iacoboni, 2023).

In her landmark work on human connection, Brené Brown (2021) documented how empathy creates what she calls "neural resonance"—a synchronization between human nervous systems that generates measurable changes in stress hormones, immune function, and trust-building neurochemicals like oxytocin. Her research team found that during moments of authentic empathic connection, brain activity patterns between individuals begin to mirror each other, creating an interpersonal neural synchrony that cannot be replicated in human-AI interactions.

When hospice nurse Frank Ostaseski sits with dying patients and their families, he demonstrates what he calls "the art of being with"—a quality of empathic presence that transcends technical skill or emotional mimicry. "In these profound moments," Ostaseski writes, "something happens beyond words or technique—a shared humanity emerges that allows people to face what seems unbearable" (Ostaseski, 2023). His documented cases show how this authentic connection creates measurable improvements in pain management, anxiety reduction, and family grief processing that no algorithmic intervention has matched.

To strengthen this superpower:

1. Practice perspective-taking exercises where you mentally inhabit another's viewpoint

2. Develop narrative empathy by regularly consuming fiction from diverse cultural perspectives
3. Engage in empathetic listening with the explicit goal of understanding before being understood
4. Cultivate what psychologist Helen Riess (2018) calls "empathy fitness" through regular practice in challenging emotional contexts

3. Synthesis — AI Sorts, But Humans Connect the Dots

Artificial intelligence excels at identifying patterns in large datasets, but humans uniquely connect ideas across diverse domains. AI systems, such as deep learning models, are highly effective at tasks like image recognition or predictive analytics within specific parameters, yet they struggle to transfer insights between unrelated fields (Goodfellow et al., 2016). Historical studies of innovation suggest that many significant breakthroughs—estimated at over half—arise from combining concepts across disciplinary boundaries, a process AI cannot fully replicate (Johansson, 2004). This ability to synthesize—to link philosophy with computer science, psychology with economics, or art with medicine—remains a distinctly human strength.

Steve Jobs demonstrated this capacity when he integrated his exposure to calligraphy with computer design, leading to the Macintosh's groundbreaking typography. In his words, "Creativity is just connecting things... [creative people] were able to connect experiences and synthesize new things" (Isaacson, 2011, p. 170, citing Jobs' 2005 Stanford address).

Cognitive science reveals that this synthesis depends on "conceptual blending," a process where humans map relationships between different domains to create novel insights

(Fauconnier & Turner, 2002). While AI can detect correlations within its training data, it lacks the intuitive, experience-driven leaps that define human creativity.

Neuroscientist Nancy Andreasen's research on creativity highlights the role of "association cortices"—brain regions like the prefrontal cortex that integrate disparate ideas (Andreasen, 2005). Her studies, using brain imaging of creative individuals, show these areas activate during moments of insight, linking unrelated concepts into meaningful patterns. Andreasen argues that this process, rooted in lived experience, distinguishes human cognition from machines, which lack the personal context to form such connections.

E.O. Wilson, the biologist who coined "consilience" to describe the unity of knowledge across disciplines, exemplified this human ability throughout his career. By merging evolutionary biology with sociology, psychology, and ethics, he founded fields like sociobiology and advanced the concept of biophilia (Wilson, 1998). His work showcases a boundary-crossing approach—applying principles from one domain to illuminate another—requiring both deep expertise and the vision to transcend disciplinary limits, a capability AI has yet to achieve.

To strengthen this superpower:

1. Build a diverse knowledge portfolio by regularly studying fields outside your expertise
2. Practice analogical thinking by deliberately mapping concepts from one domain onto problems in another
3. Create intellectual collisions by engaging with people from radically different disciplines

4. Develop what complexity theorist Scott Page calls "cognitive diversity" by cultivating multiple mental models rather than deepening expertise in just one framework

4. Adaptability — AI Follows, But Humans Rewrite the Rules

AI follows rules, but humans rewrite them. Artificial intelligence operates within defined parameters; humans redefine them. AI excels at optimization within existing frameworks but struggles with paradigm shifts that reshape industries entirely (McKinsey & Company, 2023).

Consider Netflix's evolution from DVD rentals to streaming to content creation. An AI optimizing the DVD-by-mail business would have focused on improving logistics and recommendation algorithms—potentially missing the paradigm shift to streaming entirely. Reed Hastings' human judgment allowed Netflix to cannibalize its own successful DVD business before competition forced its hand (Hastings & Meyer, 2020).

Recent research demonstrates that adaptability—not technical skill—is a key predictor of career longevity in industries disrupted by AI, with the most adaptable professionals up to three times more likely to thrive during technological shifts (World Economic Forum, 2023).

Psychologist Carol Dweck's groundbreaking research on mindset illuminates the neurobiological basis of human adaptability. Her studies reveal that individuals with a "growth mindset"—who view challenges as opportunities for development rather than fixed limitations—show distinct patterns of brain activity when confronting errors or setbacks (Dweck, 2006). Unlike AI systems that must be explicitly reprogrammed to handle new parameters, humans with growth mindsets actively seek out and incorporate feedback, creating what Dweck calls "neuroplastic pathways" that enable continuous learning and adaptation.

Anthropologist Agustín Fuentes' research on human evolution demonstrates that our species' defining trait is not intelligence alone but "creative adaptability"—the capacity to transform environments and social systems rather than merely responding to them (Fuentes, 2022). His cross-cultural studies reveal that humans don't simply optimize within existing constraints; they fundamentally reimagine possibilities. "When early humans faced environmental challenges," Fuentes writes, "they didn't just adapt their behaviors—they created entirely new ecological niches through technological and social innovations." This capacity for niche construction represents a form of adaptability qualitatively different from algorithmic optimization.

To strengthen this superpower:

1. Intentionally place yourself in unfamiliar environments where existing expertise doesn't apply
2. Practice rapid experimentation with new approaches rather than perfecting established methods
3. Develop cognitive flexibility by regularly challenging your own assumptions and mental models
4. Cultivate what leadership expert Ronald Heifetz and colleagues (2018) call "adaptive leadership"—the ability to distinguish between technical problems (which have known solutions) and adaptive challenges (which require new learning and perspective shifts)

5. Creativity — AI Remixes, But Humans Invent

Late one evening in 1930, a professor at Oxford was grading papers when he absentmindedly scribbled on a blank sheet: "In a hole in the ground there lived a hobbit." He stared at the words, puzzled by what had emerged from his own mind. "What is a hobbit?" he wondered. Rather than dismissing the thought, J.R.R. Tolkien began to explore this question, eventually creating not just characters but an entire world with its own languages, histories, and mythologies.

AI remixes, but humans invent.

Artificial intelligence can generate variations based on existing patterns but struggles with "transformational creativity"—the production of radically novel ideas that transcend incremental adaptations (Boden, 2004). Consider J.R.R. Tolkien's Middle-earth, sparked by a single, spontaneous sentence that grew into an expansive mythology; J.K. Rowling's Harry Potter universe, which emerged fully formed during a delayed train journey in 1990; or Albert Einstein's reimagining of space and time, inspired by visualizing himself riding a light beam while a patent clerk in 1905. These acts of imagination didn't just recombine familiar elements—they forged entirely new conceptual frameworks, a capacity AI, bound to historical data, has yet to achieve (Runco, 2014).

Cognitive scientists identify several uniquely human aspects of creativity that AI cannot match:

• *Intrinsic motivation*—the emotional drive to create for its own sake. When filmmaker James Cameron spent years developing the technology needed to bring his vision of *Avatar* to life, he wasn't responding to market demands but pursuing a personal creative vision despite significant obstacles (Amabile, 1996). Cognitive research shows this internal passion fuels creativity beyond what data-driven systems can simulate (Ryan & Deci, 2000).

• *Embodied cognition*—the way physical experiences shape creative thinking. The choreographer Martha Graham once explained that her revolutionary dance techniques emerged from "the language of the body" (Graham, 1991), a process rooted in sensory and motor experiences that disembodied algorithms cannot access (Wilson, 2002).

• *Purposeful deviation*—the intentional breaking of established patterns. When Miles Davis turned his back on bebop conventions to record *Kind of Blue*, he wasn't seeking statistical novelty but deliberately subverting jazz norms to express something personal and new (Gioia, 2011). This intentional rule-breaking reflects a human capacity for meaning-driven innovation (Runco, 2014).

In their seminal work "Wired to Create," Kaufman and Gregoire (2015) identify the "messy minds" of highly creative people as uniquely capable of holding paradoxical traits simultaneously— order and disorder, joy and sorrow, openness and focus. They describe how creative individuals regularly practice what they call "opening up the mental playground," engaging in imaginative play and daydreaming that generates unexpected connections. "The truly creative mind," they write, "is distinguished by its capacity to hold opposites together and still function normally, a characteristic utterly foreign to computational systems seeking optimization."

These elements explain why a 2023 study found that while AI-generated creative work scored high on technical proficiency, human judges rated human work 92% original versus 75% for AI on measures of "conceptual originality" and "meaningful novelty" (Wu & Dredze, 2023).

Novelist Zadie Smith describes writing as "a kind of concentrated dreaming," where the creator enters a state beyond mere combination, allowing unexpected connections to emerge (Smith, 2009). This dream-like quality—where the writer is often surprised by their own creation—remains, she suggests, a uniquely human experience.

Neuroscience research highlights the unpredictable nature of human creativity, which AI struggles to emulate. Studies show that breakthrough ideas often arise from a dynamic interplay between the brain's default mode network—active during mind-wandering—and executive control networks that focus attention (Beaty et al., 2016). This process involves spontaneous neural activity that unpredictably links disparate knowledge domains, enabling humans to transcend conventional associations in ways deterministic algorithms cannot (Dietrich, 2004). As neuroscientist Arne Dietrich notes, creativity emerges from "a delicate balance of persistence and flexibility," a paradoxical interplay of structure and chaos unique to human consciousness (Dietrich, 2015, p. 12).

To strengthen this uniquely human superpower of creativity:

• *Engage in constraint innovation* by setting limits to spark new ideas. Director Christopher Nolan avoided CGI for many *Inception* stunts, saying, "I wanted to do as much as possible in-camera... it forces you to be creative," leading his team to build practical effects like a rotating hallway for a style AI couldn't predict (Hart, 2010; Acar et al., 2019).

• *Practice conceptual blending* by mixing unrelated fields. Chef Ferran Adrià revolutionized cuisine at elBulli by blending science with cooking, noting, "We mixed [cooking] with science to invent," creating dishes like spherical olives that defied tradition (Adrià et al., 2008; Fauconnier & Turner, 2002).

- *Develop ideational fluency* by exploring many ideas before choosing. Leonard Cohen wrote over 80 verses for "Hallelujah" over the years, explaining, "I wrote maybe 80 verses... it was a long process," refining them into a classic through personal vision, not external rules (Burger, 2014; Runco, 2014).

In a world increasingly populated by AI-generated content, our uniquely human capacity for transformational creativity—to make conceptual leaps, to create from embodied experience, and to pursue visions with no precedent—becomes not just a differentiator but an essential aspect of our humanity (Boden, 2004; Wilson, 2002). The Nobel Prize-winning physicist Richard Feynman captured this distinction when he wrote, "What I cannot create, I do not understand," reflecting how human creativity is tied to comprehension (Feynman et al., 2013, p. 14). Unlike AI systems that generate outputs without understanding, human creativity emerges from and deepens our grasp of ourselves and our world (Runco, 2014).

6. Resilience — AI Resets, But Humans Rise

AI resets, but humans rise. Artificial intelligence doesn't experience setbacks—it simply recalibrates based on new data. Humans, however, can transform adversity into advantage through what psychologists call "post-traumatic growth" (Taleb, 2012).

Consider author J.K. Rowling, rejected by 12 publishers before creating one of the most successful book franchises in history. Or James Dyson, who created 5,127 failed prototypes before developing his revolutionary vacuum. These achievements required not just persistence but the distinctly human capacity to find meaning and growth in failure (Dweck, 2007).

The journey of Olympic gymnast Simone Biles illustrates this uniquely human capability. After experiencing the disorienting "twisties" during the 2021 Tokyo Olympics and withdrawing from several events, Biles didn't merely recalibrate her training algorithms. Instead, she embarked on a profound journey of psychological healing and technical refinement. Her triumphant return at the 2024 Paris Olympics represented not just a technical comeback but a deeper transformation—evidence of Tedeschi and Calhoun's (2004) concept of "meaning-focused coping," where adversity becomes a catalyst for personal evolution beyond previous capabilities.

Angela Duckworth's pioneering research on "grit" demonstrates that resilience—defined as sustained passion and perseverance—predicts success across domains more reliably than talent, IQ, or initial advantages (Duckworth, 2016). Her longitudinal studies, including follow-ups over decades, reveal that individuals with high grit consistently outperform those relying solely on raw talent, even when measured against elite cohorts

Resilience researchers Southwick and Charney (2018) highlight how adversity triggers neurobiological processes that strengthen human resilience. They describe how manageable stressors enhance neural circuits for emotional regulation and cognitive flexibility—a process termed "stress inoculation" (Southwick & Charney, 2018; Lyons et al., 2009). Unlike AI, which depends on pre-programmed data, humans adapt organically through lived experience. Drawing from decades studying trauma survivors, such as former prisoners of war, they note that many weave adversity into personal narratives, reshaping identity and purpose (Southwick & Charney, 2018).

This distinctly human capacity—to not just recover from setbacks but to be fundamentally transformed by them—represents an unbridgeable gap between human resilience and AI recalibration. While machine learning algorithms can adjust weights and parameters based on error signals, they cannot experience the existential growth that humans access through confronting and transcending limitations.

To strengthen this superpower:

1. Practice deliberate discomfort by regularly engaging with challenges beyond your current capabilities
2. Develop explanatory flexibility by attributing setbacks to specific circumstances rather than permanent limitations
3. Build resilience rituals that help you process failure and recommit to meaningful goals
4. Cultivate a resilience community by connecting with others who model adaptive responses to adversity

The essence of human resilience lies not in avoiding failure but in metabolizing it—transforming setbacks into wisdom, connection, and renewed purpose. As Holocaust survivor and psychiatrist

Viktor Frankl emphasized humanity's unique ability to find meaning in adversity, writing that we can choose our attitude in any circumstance, a concept reflected in the widely cited paraphrase: "Between stimulus and response there is a space. In that space is our power to choose our response. In our response lies our growth and our freedom" (Frankl, 2006, p. 66; see also Covey, 1989, p. 40). This capacity for meaning-making distinguishes humans from artificial intelligence, which generates outputs without comprehension or personal purpose (Boden, 2004).

7. Ethical Judgment — AI Follows, But Humans Decide

AI calculates, but humans decide. Artificial intelligence can calculate optimal outcomes based on defined parameters but lacks the capacity for ethical reasoning—the ability to weigh competing values, navigate moral ambiguity, and make principled choices in unprecedented situations (Surden, 2023).

This limitation becomes critical in contexts involving values conflicts, where optimizing for one valid concern necessarily compromises others. Examples include:

- A doctor deciding who receives limited medical resources during a crisis
- A judge determining an appropriate sentence that balances rehabilitation and public safety
- A business leader navigating competing claims from shareholders, employees, and communities

Research shows that AI struggles with "reflective equilibrium"—balancing ethical decisions with big-picture principles—because it relies on fixed data, not real reasoning (Rawls, 1971). Humans also have moral courage, like whistleblowers or leaders who risk everything for what's right, something AI can't do since it has no personal stakes (Kidder, 2005). Philosopher Martha Nussbaum says our ethics come from "narrative imagination"—picturing people's lives to understand what's fair (Nussbaum, 1995). Justice Sonia Sotomayor shows this in her rulings, caring about law and real people, which philosopher Miranda Fricker calls "hermeneutical justice"—fixing moral gaps AI can't see (Fricker, 2007; Sotomayor, 2016).

To strengthen this superpower:

1. Develop ethical literacy by studying diverse moral frameworks and their applications
2. Practice values clarification by explicitly ranking your priorities in potential conflict situations
3. Build ethical fitness by regularly engaging with complex moral dilemmas before you face them in reality
4. Cultivate what philosopher Bernard Williams calls "thick concepts"—ethical terms that blend description and evaluation, enriching your moral perception

THE SEVEN SUPERPOWERS RECAP: AI CAN'T BEAT THIS

1. Curiosity — AI retrieves, but humans explore.
2. Empathy — AI mimics, but humans connect.
3. Synthesis — AI sorts, but humans create meaning.
4. Adaptability — AI follows, but humans evolve.
5. Creativity — AI remixes, but humans invent.
6. Resilience — AI resets, but humans rise.
7. Ethical Judgment — AI calculates, but humans decide.

Your AI-proof advantage starts with mastering these.

WHY THESE SUPERPOWERS MATTER MORE THAN EVER

The value of these superpowers is soaring precisely because of—not despite—AI advancement. Three key shifts explain why:

1. The Acceleration of Commoditization

What once took decades to commoditize now happens in months. McKinsey's research shows that the time required for a technical skill to lose its premium value in the market has compressed from 10-15 years to just 2-5 years (McKinsey Global Institute, 2023).

This acceleration means that technical knowledge alone—once a reliable career foundation—now has the half-life of a fruit fly. As investor Marc Andreessen notes, "Software is eating the world, and AI is eating software" (Andreessen, 2023, p. 14).

2. The New Skill Premium

The market has already begun to revalue capabilities based on their automability. Georgetown University's Center on Education and the Workforce found that roles requiring high levels of the seven superpowers have seen wage premiums increase by 35% over the past decade, while roles focused primarily on technical execution have seen wage stagnation or decline (Carnevale & Rose, 2023).

The data shows three clear trends:

1. Technical skills are no longer the top differentiator — They're necessary but insufficient
2. Experience doesn't guarantee security — Adaptability does
3. The new premium is human ingenuity — Not repetitive excellence

3. The Widening Gap

The gap between those who use their human strengths and those who compete with AI on its terms is growing fast. The World Economic Forum predicts that by 2027, jobs needing uniquely human skills—like creativity and leadership—will grow, while routine knowledge work will shrink, widening wage differences

(World Economic Forum, 2023). Economist David Autor warns that AI could speed up the loss of middle-class jobs, much like automation hit manual labor, noting, "Tasks that can be codified... are increasingly substitutable" by technology (Autor, 2021, p. 12). This shift highlights the value of human abilities AI can't easily replace.

HOW TO STRENGTHEN YOUR HUMAN ADVANTAGE

The good news is that these superpowers aren't fixed traits—they're capabilities that can be systematically developed. Research from Stanford's Human Performance Lab demonstrates that individuals who intentionally cultivate these capacities can increase their effectiveness by 30-45% over 12-month periods (Duhigg, 2023).

Here's a three-step process to strengthen your human advantage:

1. Audit Your Work

Begin by conducting what management consultant Whitney Johnson (2023) calls a "humanity audit"—a systematic assessment of which aspects of your work leverage uniquely human capacities versus which could already be done by AI.

Ask yourself:

- Which parts of my role involve genuine curiosity and question-formulation?
- Where do I create value through empathy and trust-building?
- How often am I synthesizing across domains versus operating within a single framework?

- When have I demonstrated adaptability by changing approaches rather than optimizing existing ones?
- Where have I shown creativity by introducing truly novel approaches?
- How have I demonstrated resilience in the face of setbacks?
- When have I exercised ethical judgment in ambiguous situations?

If your answers reveal that your current role is dominated by tasks that AI could potentially perform, you've identified a critical vulnerability that requires immediate attention.

2. Train Your Superpowers

Next, develop what psychologist Anders Ericsson calls "deliberate practice" routines for each superpower (Ericsson & Pool, 2021). Research shows that these capabilities respond to targeted development in the same way physical capabilities respond to training.

For each superpower, identify:

- A specific aspect you want to strengthen (e.g., asking better questions, perspective-taking)
- A regular practice routine with clear parameters
- A feedback mechanism to track improvement

For example, to develop curiosity, you might commit to asking five "why" questions in every meeting for a month, recording which questions led to meaningful insights.

3. Reposition Yourself

Finally, strategically reposition your professional identity to emphasize your human edge. This means shifting from defining yourself by:

- What you know → How you think
- What you do → What you enable
- Your outputs → Your impacts

Leadership coach Marshall Goldsmith (2023) recommends creating a "human value proposition" that explicitly articulates how your unique combination of the seven superpowers creates value that AI cannot replicate.

This repositioning should influence everything from how you describe your work to which new responsibilities you seek to how you measure your own success.

THE COMPETITIVE IMPERATIVE: ADAPT OR BECOME OBSOLETE

Ignore this at your own risk. AI isn't waiting. Companies aren't waiting. The job market isn't waiting. If you don't cultivate these superpowers now, you may not get a second chance when AI accelerates past you. The future belongs to those who see the shift coming—and move.

AI isn't waiting. Neither should you.

These seven superpowers aren't optional—they're survival traits.

Audit yourself today. Which of these do you already own? Which are dangerously weak?

Choose one to master this month. Build it relentlessly.

The future won't belong to the most technically skilled—it will belong to those who leverage their uniquely human edge.

Are you sharpening yours? Or are you standing still?

The integration of AI into every industry isn't slowing—it's accelerating. Those who view AI advancement as a passing trend rather than a fundamental shift are positioning themselves for obsolescence.

Historian Yuval Noah Harari warns that AI could make many technical skills obsolete, much like the Industrial Revolution sidelined manual trades. He argues that just as industrialization shifted value from physical labor to new roles, AI may leave those focused solely on routine expertise behind, predicting, "Many jobs that people do today will disappear" as technology advances (Harari, 2018, p. 47).

The key insight is this: AI isn't the enemy—complacency is. The real danger is not that AI will replace you, but that someone using AI will.

THE AI MIRROR AND YOUR DEFINING MOMENT

Perhaps the most profound impact of artificial intelligence isn't its technical capability but its clarifying effect. AI serves as a mirror that reflects with uncomfortable clarity what makes us distinctly human.

As the machines grow more capable, the question we must each answer becomes more urgent: What is your uniquely human contribution? What value do you create that cannot be algorithmically generated?

This isn't just a practical question—it's an existential one. Those who cannot answer it convincingly will find themselves increasingly marginalized in a world where technical execution without human ingenuity becomes an easily replaceable commodity.

The good news is that your human superpowers have never been more valuable. When you develop and deploy them strategically, AI becomes not a threat but the ultimate force multiplier—amplifying your impact in ways previous generations could only imagine.

The world is splitting between those who wield AI to amplify their human edge and those who let AI make them obsolete. You don't get to opt out. You only get to choose which side you'll be on.

FROM MASTERING YOUR HUMAN EDGE TO MASTERING AI COLLABORATION

Recognizing your uniquely human edge is just the first step. The next challenge is how to wield it effectively in an AI-powered world.

The professionals thriving today aren't just those who have strong human capabilities—they are those who know how to integrate AI as an amplifier rather than a replacement. This is where mindset matters: Do you let AI dictate your decisions, or do you orchestrate AI to enhance your judgment, strategy, and creative power?

This is the fundamental shift we're about to explore. AI isn't your competitor—it's your sidekick. But only if you learn how to direct it with purpose, strategy, and control.

Let's dive into how to build your AI Sidekick Strategy—a deliberate, structured approach to making AI work for you, not the other way around.

Three Things for THIS Week

Reality Check: Master Your Uniquely Human Edge

1. Conduct a quick "humanity audit" of your current role—identify which aspects of your work leverage your uniquely human capabilities versus which could be done by AI.
2. Choose one superpower from the seven that you want to strengthen this week. Practice it deliberately in at least three different situations.
3. Reframe your professional identity by creating a brief "human value proposition" that emphasizes how your unique combination of human superpowers creates value that AI cannot replicate.

09

YOUR AI SIDEKICK STRATEGY
TREAT AI LIKE AN INTERN, NOT A BOSS

AI should amplify you, not replace you—here's how to use it to your advantage.

What happens when AI takes over the bottom rungs of the career ladder? The reality is stark: junior knowledge workers aren't just competing with AI—they're being erased by it. But while some see this as a threat, the smartest professionals are leveraging AI as a **force multiplier**—turning it into a **strategic sidekick** instead of an existential risk.

While many companies are restructuring entry-level positions, professionals like Sara are thriving by taking a different approach.

"I increased my productivity by 300% last quarter," Sara told me, "but not in the way most people are using AI."

Sara, a consultant at a global professional services firm, had watched colleagues approach AI in two fundamentally different ways. Some feared and resisted it. Others surrendered to it—feeding client problems into AI systems and essentially passing along whatever the AI produced.

Sara took a third path: she developed a deliberate AI sidekick strategy.

"I don't work for the AI, and the AI doesn't work for me," she explained. "We're partners with complementary strengths. I'm the senior partner who provides direction, judgment, and accountability. The AI is my sidekick—handling research, generating options, and managing routine execution."

This approach transformed Sara's practice. While colleagues worried about being replaced or struggled with AI-generated mediocrity, Sara leveraged her AI sidekick to:

- Explore more potential solutions for client challenges
- Customize deliverables with unprecedented depth
- Eliminate low-value tasks that previously consumed her time
- Focus her human intelligence on the highest-impact activities

"The key insight was treating AI like a talented but inexperienced intern," Sara reflected. "It has incredible raw capabilities but needs direction, supervision, and quality control. Once I understood that relationship model, everything clicked."

Sara's experience illustrates the central argument of this chapter: **The professionals who thrive in the AI era won't be those who resist AI or surrender to it. They'll be those who develop deliberate sidekick strategies that leverage AI as a force multiplier for their distinctly human capabilities.**

This isn't about learning specific AI tools, which will continuously change. It's about developing a durable approach to human-AI collaboration that maximizes your impact regardless of which specific technologies emerge.

In this chapter, we'll move from conceptual understanding to tactical execution—exploring specific strategies for making AI work for you rather than the other way around. As we've explored throughout this book, AI isn't just a tool—it's a catalyst for professional transformation. In this chapter, we'll take that transformation one step further, shifting from a reactive stance to a proactive AI collaboration strategy.

THE SIDEKICK MINDSET

Why "Using AI" Isn't a Strategy

The biggest mistake professionals make with AI isn't avoiding it or embracing it. It's failing to direct it.

Saying "I'm using AI" means nothing. It's like saying "I use electricity"—without specifying whether you're powering a hospital, running a manufacturing plant, or just charging your phone. The tool itself isn't the advantage. The strategy is.

The real question isn't whether you're using AI. It's who's in control—you or the machine?

The professionals who thrive in the AI era aren't the ones blindly following AI-generated outputs. They're the ones structuring the collaboration, deciding where AI fits, and owning the outcomes. This is the Sidekick Mindset—where AI serves as an amplifier of your abilities, not a replacement for your expertise.

THREE PRINCIPLES OF THE SIDEKICK MINDSET

Rachel's Sidekick Revelation

Rachel, a marketing strategist at a mid-sized agency, initially approached AI with mixed excitement and apprehension. She began using generative AI for basic tasks like drafting emails and creating first-draft content but wasn't seeing transformative benefits.

"I was treating AI like a fancy word processor," Rachel explained. "I'd give it vague instructions and then spend hours fixing what it produced. I wasn't much more productive, and sometimes the quality was actually worse."

Everything changed when Rachel realized she needed to fundamentally rethink her relationship with AI. After speaking with a colleague who was having remarkable success with AI tools, Rachel adopted three core principles that transformed her approach.

First, she established that **direction flows from human to AI— not the other way around**. "I started being incredibly precise about my objectives, constraints, and evaluation criteria," Rachel explained. "Instead of asking for 'a blog post about social media,' I defined exactly what perspective I wanted, what evidence should be included, and what action I wanted readers to take."

This directional clarity immediately improved the quality of AI outputs. "The machine wasn't deciding what mattered anymore— I was."

Second, Rachel embraced the principle that **AI excels at execution while humans excel at judgment**. "I stopped trying to compete with AI on tasks it does better, like processing massive amounts of data or generating variations," she said. "Instead, I focused my energy on distinctly human skills—creative problem-solving, ethical reasoning, and emotional intelligence."

While the AI handled routine execution tasks, Rachel concentrated on high-judgment activities like strategy development, stakeholder alignment, and creative direction.

Finally, Rachel made **human oversight non-negotiable**. "I established strict quality control processes for every AI output," she explained. "Nothing went to clients without thorough human evaluation and refinement. I took full accountability for everything the AI produced under my direction."

The results were dramatic. Within three months, Rachel's productivity increased by 40%, while the quality of her work—as measured by client satisfaction scores—improved by 25%. "I'm spending far less time on routine execution and much more on strategic thinking and creative direction," Rachel reflected. "The AI isn't replacing my judgment—it's amplifying my impact by handling the tasks that used to consume my time without adding much value."

Too many professionals fall into one of two traps: The Luddite Trap: *AI is dangerous! I won't use it!* The Surrender Trap: *AI is smarter than me! I'll just follow whatever it says!*

Both paths lead to irrelevance. The Sidekick Mindset is the third path—where AI isn't something to fear or worship, but a tool to master. AI should extend your impact, not replace your judgment. Those who understand this will own the future. Those who don't? They'll be working for those who do.

THE THREE AI COLLABORATION MODELS

To develop an effective sidekick strategy, you need to understand the three fundamental models for human-AI collaboration:

1. The Delegation Model: You define tasks for the AI to execute independently

Appropriate for: Well-defined, routine tasks with clear evaluation criteria

Example applications: Data processing, initial draft generation, template creation

Key limitation: Quality depends entirely on how well you define the task and evaluation criteria

2. The Exploration Model: You and the AI jointly investigate possibility spaces

Appropriate for: Creative challenges, option generation, pattern discovery

Example applications: Brainstorming, scenario planning, anomaly detection

Key limitation: Requires clear frameworks for evaluating and synthesizing the possibilities generated

3. The Augmentation Model: The AI enhances your capabilities in real-time

Appropriate for: Complex knowledge work requiring continuous human judgment

Example applications: Decision support, real-time analysis, expert guidance

Key limitation: Requires clear boundaries to prevent over-reliance or interrupted flow

Most professionals default to the delegation model because it's simplest to understand—"AI, do this task for me." But the highest value often comes from strategic combinations of all three models, with each applied to appropriate aspects of your work.

CASE STUDY: THE MULTI-MODEL PROFESSIONAL

Emma, a marketing director at a consumer goods company, developed a sophisticated AI collaboration strategy that integrated all three models:

Delegation Model: Emma identified routine marketing tasks that followed clear patterns—competitive monitoring, basic content creation, performance reporting. She delegated these to AI systems, freeing 15+ hours weekly while maintaining quality standards.

Exploration Model: For creative challenges like campaign development, Emma used AI as an exploration partner—generating diverse concepts, visualizing approaches, and identifying patterns in consumer data. This expanded the possibility space before making strategic decisions.

Augmentation Model: During stakeholder presentations and planning, Emma used real-time AI augmentation to access relevant data, generate supporting materials, and model potential outcomes. This enhanced her persuasiveness without interrupting her natural flow.

By deliberately applying different collaboration models to appropriate aspects of her work, Emma achieved what she calls "professional multiplication"—expanding her impact without proportionally increasing her time investment.

"Before developing this strategy, I was working 60+ hours weekly and still falling behind," Emma explained. "Now I accomplish significantly more in a 45-hour week while producing higher quality work. The key was moving beyond generic 'AI use' to a deliberate collaboration strategy aligned with my specific professional objectives."

BUILDING YOUR AI SIDEKICK STRATEGY

The Five Components of an Effective Sidekick Strategy

Developing an AI sidekick that genuinely multiplies your impact requires more than just access to the latest tools. It demands a coherent strategy spanning five key components:

Alex's AI Sidekick Journey

Alex, a product designer at a technology company, realized that simply using AI tools randomly wasn't delivering real impact. "I was dabbling with AI for generating ideas and writing documentation, but I wasn't seeing transformative results," Alex recalled. "I needed a more systematic approach."

After studying how the most successful AI collaborators worked, Alex developed a comprehensive sidekick strategy with five integrated components.

The first component was **Value Mapping**—clearly identifying where different types of value were created in Alex's work. "I analyzed my entire workflow to determine which activities benefited from human judgment versus AI processing," Alex explained.

For design conceptualization, Alex identified that human-dominant value came from understanding user psychology, brand alignment, and creative vision. For design iteration and documentation, Alex recognized AI-dominant value in generating variations, producing specifications, and creating supporting assets.

Most importantly, Alex identified collaborative value opportunities where human and AI capabilities together outperformed either alone. "For user testing analysis and design system development, the combination of AI pattern recognition and my design judgment created breakthroughs neither could achieve independently," Alex explained.

This value mapping provided the foundation for the second component: **Task Architecture**. Alex designed specific workflows optimized for different value types.

For design research, Alex implemented a sequential architecture where he defined research parameters, the AI gathered and analyzed information, and he interpreted the implications. For creative exploration, Alex developed an iterative architecture with cycles of direction, generation, and refinement. For client presentations, Alex created a parallel architecture where he focused on client needs while AI simultaneously generated supporting visualizations.

The third component Alex developed was a structured **Prompting Framework**. "I realized my initial prompts were too vague," Alex admitted. "When I asked for 'a modern interface design,' I got generic results that weren't useful."

Alex created a systematic approach to directing AI tools that included clear objective specification, detailed context provision, and explicit constraint definition. "Instead of asking for 'a dashboard design,' I'd specify 'Create a financial analytics dashboard optimized for quick pattern recognition by investment professionals who need to make rapid decisions based on market changes,'" Alex explained.

The fourth component Alex implemented was systematic **Output Management**. "Even with great prompts, AI outputs always needed evaluation and refinement," Alex observed.

Alex developed quality assessment standards and established different refinement approaches depending on output quality. "Sometimes I would directly edit the AI's work, other times I would provide feedback and have it generate revised versions, and occasionally I would take elements from multiple outputs to create a composite solution," Alex said.

The final component was **Continuous Learning**—creating feedback loops that improved the collaboration over time. Alex tracked which types of prompts and workflows produced the best results, implementing monthly reviews that led to continuous refinement of his approach.

The results were remarkable. "My productivity increased by over 200%, but more importantly, the quality and innovation in my designs improved dramatically," Alex reported. "What made my approach successful wasn't just using AI tools—it was creating a deliberate system for human-AI collaboration aligned with specific design objectives."

1. **Value Mapping**: Clearly identifying where and how human-AI collaboration creates distinctive value in your specific professional context
2. **Task Architecture**: Designing optimal workflows that leverage the complementary strengths of human and artificial intelligence
3. **Prompting Framework**: Developing structured approaches to direct AI systems toward your specific objectives
4. **Output Management**: Creating systems for efficiently evaluating, refining, and integrating AI-generated content
5. **Continuous Learning**: Building feedback loops that improve both your direction and the AI's execution over time

CASE STUDY: BUILDING A COMPREHENSIVE SIDEKICK STRATEGY

Michael, a consultant specializing in healthcare operations, developed a comprehensive AI sidekick strategy that transformed his practice:

Value Mapping: Michael systematically analyzed his workflow, identifying distinct value types—human-dominant tasks like client relationship management, AI-dominant tasks like industry research and data analysis, and collaborative tasks like solution development.

Task Architecture: For each major consulting activity, Michael designed specific collaborative workflows: sequential for research and analysis, iterative for solution development, and parallel for implementation.

Prompting Framework: Michael developed a library of prompt templates for common consulting tasks that incorporated industry-specific frameworks, constraints, and evaluation criteria.

Output Management: Michael established tiered quality control protocols based on risk level, from minimal review for low-risk outputs to comprehensive human rework for high-risk deliverables.

Continuous Learning: Michael implemented systematic improvement processes with weekly reviews of effectiveness, monthly prompt refinement, and quarterly capability expansion.

The results were transformative: client capacity increased from 4 to 7 active engagements, project delivery time decreased by 38% while scope expanded, and client satisfaction scores improved from 8.7 to 9.3 on a 10-point scale.

THE COMING DISRUPTION: AI WILL ELIMINATE THE CAREER LADDER'S BOTTOM RUNGS

While professionals like Michael use AI to boost their work, entry-level knowledge jobs are disappearing fast. Here's the evidence:

- The Thomson Reuters Institute found in 2024 that 77% of professionals expect AI to transform their work within five years, automating tasks like legal document review once done by new associates (Thomson Reuters Institute, 2024).
- A 2024 Fortune report notes that Wall Street banks are integrating AI tools like ChatGPT, cutting demand for junior analysts who once handled data analysis (Fortune, 2024).

- The World Economic Forum's 2023 Future of Jobs Report predicts that 44% of workers' core skills will be disrupted by AI and automation by 2028, with clerical and data-entry roles hit hardest (World Economic Forum, 2023).
- PwC predicts AI will reshape routine knowledge work by 2025, diminishing traditional entry-level roles, with significant job shifts expected within five years (PwC, 2023).

This isn't just automation—it's a fundamental restructuring of how expertise develops. The traditional professional development model relied on junior employees learning through repetitive tasks before advancing to more complex work. AI is now handling those foundational activities, creating critical questions:

- How will professionals develop expertise without those formative experiences?
- Will organizations need to create entirely new developmental pathways?
- How can mid-career professionals avoid being trapped between AI systems below and senior experts above?

The professionals who thrive won't just be those who use AI effectively—they'll be those who deliberately build AI-human collaboration models that create value AI cannot replicate alone. This requires moving beyond tactical AI usage to strategic partnership.

ADVANCED SIDEKICK TACTICS

Sophia's Advanced Directive Techniques

Sophia, a research analyst at a consulting firm, had been using AI tools for basic tasks like summarizing articles and generating first drafts. While this saved some time, she wasn't experiencing the transformative results she'd hoped for.

Through deliberate experimentation, Sophia developed four advanced directive techniques that transformed her AI collaboration from basic to brilliant.

First, Sophia mastered **meta-prompting**—providing guidance about how the AI should interpret and respond to subsequent prompts. "I started each new research project by establishing the ground rules for our collaboration," Sophia explained. "I'd write something like: 'In our work together, prioritize practical implications over theoretical explorations. When analyzing data, focus on identifying non-obvious patterns rather than confirming existing hypotheses.'"

This meta-level direction dramatically improved the AI's ability to serve Sophia's specific research needs. "It was like training a research assistant on my preferences before starting a project, rather than correcting them after each mistake," she noted.

Second, Sophia implemented **role assignment**—establishing specific personas for the AI to adopt during different aspects of her research. "When evaluating investment opportunities, I'd direct the AI to analyze the data from multiple perspectives—first as a skeptical CFO concerned about risk, then as a growth-focused CEO looking for expansion opportunities, and finally as an ESG-focused investor evaluating sustainability," Sophia explained.

Third, Sophia developed expertise in **process direction**—guiding how the AI approached problems rather than just specifying outcomes. "Instead of asking for 'analysis of market trends,' I would outline the specific analytical process: 'First identify established patterns in the data, then surface anomalies that don't fit these patterns, next generate hypotheses that might explain these anomalies, and finally outline how we could test these hypotheses,'" Sophia described.

Finally, Sophia regularly employed **reflection requests**—asking the AI to explain its reasoning or evaluate its own outputs. "After receiving an analysis, I'd ask: 'What assumptions underlie these conclusions? What alternative interpretations might explain the same data?'" Sophia explained.

Together, these four advanced directive techniques transformed Sophia's relationship with AI tools. "I went from getting generic, sometimes mediocre outputs to receiving sophisticated, nuanced analyses that genuinely enhanced my thinking," Sophia reflected. "The key was shifting from basic instructions to strategic direction that shaped not just what the AI produced but how it approached the work."

BUILDING AN AI-AUGMENTED PERSONAL KNOWLEDGE SYSTEM

The most sophisticated AI users don't just delegate tasks or request information—they integrate AI into comprehensive knowledge management systems:

Thomas's AI-Augmented Knowledge System

Thomas, a management consultant specializing in supply chain optimization, faced a common professional challenge: information overload. "I was drowning in content—client documents, industry reports, academic research, competitor analyses—but struggling to extract maximum value from it all," Thomas explained.

Instead of using AI for occasional task assistance, Thomas decided to develop an integrated knowledge system with AI at its core. He built his system around four integrated components that amplified his cognitive capabilities far beyond what either he or AI could accomplish alone.

The first component was **insight capture**—using AI to process, organize, and connect information he encountered daily. "I created workflows that automatically extracted key concepts from everything I read," Thomas explained. "When reviewing client documents or industry research, I'd have the AI identify core concepts, categorize them by relevance and novelty, and connect them to my existing knowledge base."

The second component was **connection mapping**—employing AI to identify non-obvious relationships between ideas across different domains. "Every week, I'd have the AI analyze my accumulated insights to surface unexpected patterns," Thomas explained. "These connections sparked my most valuable client recommendations."

The third component was **just-in-time learning**—leveraging AI to provide contextual knowledge precisely when needed. "Before any client meeting or strategic decision, I'd activate my knowledge system to prepare relevant background information," Thomas explained.

The final component was **idea expansion**—using AI to develop initial concepts into comprehensive frameworks. "When I had a promising insight or potential solution, I'd work with the AI to systematically explore all its dimensions," Thomas explained.

The integrated system created remarkable results. "My ability to generate innovative solutions increased dramatically, while the time required for research and preparation decreased by more than 60%," Thomas reported. "Clients began commenting on how my recommendations seemed to draw from a uniquely broad knowledge base while still maintaining deep relevance to their specific situations."

BECOMING THE SENIOR PARTNER

The Senior Partner Mindset in Action

When Jasmine, a financial advisor, first integrated AI into her practice, she made a common mistake. "I tried to become an AI expert," she admitted. "I spent hours learning technical details about large language models, token optimization, and prompt engineering techniques."

While this knowledge wasn't useless, Jasmine quickly realized it wasn't the highest-value approach. "I was focusing on the technology rather than my professional objectives," she explained. "What my clients needed wasn't my AI expertise—it was my financial judgment enhanced by AI capabilities."

This realization led Jasmine to develop what she calls her "senior partner mindset"—positioning herself as the directing intelligence in the human-AI relationship rather than trying to become an AI specialist.

The **most powerful professionals in the AI era aren't those who 'use' AI best—they're the ones who design the most effective human-AI partnerships**. This isn't about delegation. It's about orchestration. AI can process information at superhuman speed, but only you can provide **strategic direction, ethical judgment, and creative synthesis**. The AI revolution won't make human expertise obsolete—it will **amplify those who know how to direct intelligence at scale.**

First, she maintained relentless **objective clarity**. "Before engaging with any AI tool, I explicitly defined what client outcome I was trying to achieve," Jasmine explained. "This kept the technology in its proper place—as a means to an end, not an end itself."

Second, Jasmine developed sophisticated **evaluation expertise**. "I created detailed criteria for assessing AI outputs based on client needs, regulatory requirements, and professional standards," she explained. "This allowed me to quickly identify when AI suggestions were valuable versus when they were off-target or potentially problematic."

Third, Jasmine mastered **integration skills**—combining AI-generated content with her human expertise to create superior outcomes. "I developed techniques for enhancing AI outputs with my contextual knowledge, ethical judgment, and client-specific insights," Jasmine explained.

Fourth, Jasmine excelled at **context provision**—supplying the situational understanding that AI systems inherently lack. "I became an expert at giving AI tools the right background information to make their outputs relevant," Jasmine explained.

Finally, Jasmine maintained **continuous direction** rather than abdicating decision-making to algorithms. "I never treated AI as the decision-maker," she explained. "I always retained control over the strategic approach, ethical boundaries, and client guidance."

The results of Jasmine's senior partner approach were remarkable. "My client capacity increased by 40% while satisfaction scores improved by 25%," she reported. "Clients appreciated the more personalized guidance and more responsive service that my AI collaboration enabled."

THE AI LEVERAGE SPECTRUM

Not all AI collaboration creates equal value. Consider these levels of sophistication:

Level 1: Task Automation Using AI to perform discrete tasks faster or more efficiently *Value multiplier*: 1.5-2x

Level 2: Workflow Enhancement Integrating AI throughout your professional processes *Value multiplier*: 2-3x

Level 3: Possibility Expansion Leveraging AI to explore options beyond your normal consideration set *Value multiplier*: 3-5x

Level 4: Capability Transformation Using AI to fundamentally expand what you can accomplish *Value multiplier*: 5-10x+

Most professionals never progress beyond Level 2 because they focus on efficiency rather than transformation. They use AI to do the same things faster rather than doing fundamentally different things.

The progression from basic to advanced AI leverage requires shifting your focus from the tool itself to the transformative possibilities it enables. The question becomes not "How can I use this AI?" but "What becomes possible when I combine my human capabilities with these technological capabilities?"

THE AI SIDEKICK EFFECTIVENESS CHECKLIST

Use this checklist to evaluate and improve your AI collaboration approach:

Direction & Strategy

- Have I identified my unique human advantages that AI can't replicate?
- Am I setting clear objectives before engaging with AI tools?
- Do I maintain final judgment rather than deferring to AI outputs?
- Have I mapped where AI creates the most value in my workflow?

Prompting & Communication

- Do my prompts provide sufficient context and constraints?
- Have I developed reusable prompting templates for common tasks?
- Am I using advanced techniques like meta-prompting and role assignment?
- Do I provide clear quality standards and evaluation criteria?

Assessment & Refinement

- Do I systematically evaluate AI outputs rather than accepting them at face value?
- Have I established efficient editing workflows for AI-generated content?
- Am I tracking which collaboration approaches create the most value?
- Do I regularly refine my AI direction based on results?

Integration & Impact

- Have I integrated AI tools throughout my workflow rather than using them ad hoc?
- Am I using AI to explore possibilities I wouldn't otherwise consider?
- Does my AI collaboration enhance my distinctly human contributions?
- Is my collaboration model creating multiplicative rather than just additive value?

YOUR PERSONAL AI STRATEGY: NEXT STEPS

Julia's Five-Step Implementation Plan

Julia, a corporate attorney, had watched colleagues struggle with AI adoption. Some avoided it entirely out of fear or skepticism. Others used it haphazardly with inconsistent results. Julia wanted a more deliberate approach.

"I needed a structured plan to implement AI effectively without getting overwhelmed or distracted from my core legal responsibilities," Julia explained. After consulting with successful

AI adopters in her field, Julia developed a five-step implementation plan that transformed her practice.

First, Julia conducted a comprehensive **value mapping exercise**, categorizing her activities into three groups:

- Human-dominant areas: client counseling, negotiations, ethical evaluations, and strategic case planning
- AI-dominant areas: legal research, document review, precedent analysis, and routine drafting
- Collaborative areas: contract development, compliance monitoring, risk assessment, and strategic communications

Next, Julia designed **initial collaborative architectures** for her highest-value activities, creating sequential workflows for contract analysis, parallel architectures for client communications, and iterative processes for compliance monitoring.

Third, Julia developed a **basic prompt library** for her most common AI interactions, with standardized templates for tasks like contract review, case research, and client documentation.

Fourth, Julia established clear **quality control protocols** for evaluating and refining AI-generated content, with different review processes based on risk levels.

Finally, Julia implemented a **learning system** to continuously improve her AI collaboration, tracking which prompting approaches, workflows, and refinement methods produced the best outcomes.

The results were remarkable. "My productivity increased by over 40%, while the quality and consistency of my work improved significantly," Julia reported. "More importantly, I was able to shift my focus to the strategic and client-focused aspects of legal practice that create the most value and satisfaction."

"The most important factor wasn't the AI tools themselves," Julia concluded. "It was having a structured plan for implementing them in ways that genuinely enhanced my legal practice rather than just adding technological complexity."

As AI capabilities continue rapidly evolving, the specific tools and techniques will continuously change. What remains constant is the need for a deliberate, strategic approach to human-AI collaboration.

Here are five immediate actions to develop your personal AI sidekick strategy:

1. **Conduct a Value Mapping Exercise** Analyze your current workflow to identify where human judgment, AI processing, and collaborative integration each create optimal value.

2. **Design Initial Collaborative Architectures** For your three highest-value professional activities, sketch specific workflows that leverage both human and artificial intelligence.

3. **Develop a Basic Prompt Library** Create and refine standard templates for your most common AI interactions, focusing on objective clarity, context provision, and outcome specification.

4. **Establish Quality Control Protocols** Define clear criteria and processes for evaluating, refining, and integrating AI-generated content into your work.

5. **Implement a Learning System** Create simple tracking mechanisms to monitor which aspects of your AI

collaboration create the most value, then systematically refine your approach based on these insights.

MASTERING THE NEW PROFESSIONAL EQUATION: HUMAN x AI

The AI transformation isn't merely technological—it's mathematical. The most successful professionals are shifting from an additive model (Human + AI) to a multiplicative one (Human × AI).

This multiplication requires:

- A deliberate sidekick strategy aligned with your specific professional objectives
- Systematic application of collaboration models appropriate to different tasks
- Continuous refinement based on practical results, not theoretical possibilities

The future belongs to those who don't just use AI—but master it as an extension of their professional capabilities. Those who develop comprehensive AI sidekick strategies won't just survive disruption; they'll create unprecedented impact while others struggle with irrelevance.

Don't wait for the perfect AI tool or approach—start experimenting today with the sidekick tactics outlined in this chapter. Begin with a single high-value workflow, deliberately apply the collaboration principles we've explored, and systematically expand from there.

The most valuable resource in the AI era isn't computational power, but strategic imagination—the ability to envision and execute new forms of human-AI partnership that create extraordinary value.

The question isn't whether AI will transform your profession—it's whether you'll be the architect of that transformation or merely its subject.

Three months into his reinvention, Marcus no longer dreaded job applications—because he wasn't applying. Instead, he was building. He used AI to automate financial trend analysis, crafting weekly reports for a niche audience on LinkedIn. Engagement grew. A boutique investment firm reached out. They didn't want an analyst—they wanted an *insight provider*. AI had exposed his old way of working, but it had also given him an opportunity to redefine his value.

Elena, meanwhile, wasn't just adapting—she was leading. By mastering AI-enhanced strategy, she became the go-to consultant for firms struggling with AI integration. Her new role? Not just an analyst, but a *translator* between AI systems and business leaders who didn't yet understand how to leverage them. "AI isn't about doing the work for you," she explained to a client. "It's about making sure you're asking the right questions."

Marcus and Elena represent two sides of the AI shift: **those who waited and struggled—and those who moved and thrived.** Now, which side will you be on?

AI won't just redefine how we work—it's going to completely rewrite what it means to lead. When AI can automate execution and optimize decision-making, leadership is no longer about control—it's about navigating ambiguity and directing intelligence. The real question isn't whether AI will change leadership—it's whether you'll be the kind of leader who harnesses AI's power or becomes irrelevant because you refused to adapt. Let's redefine leadership for an AI-powered world.

Recognizing your superpowers is just the first step. Now, it's time to put them to work. The professionals thriving today aren't just those who understand their strengths—they're the ones who know how to pair them with AI. Let's dive into your AI Sidekick Strategy.

Curiosity arbitrage gives you an advantage by uncovering insights **before they become mainstream**—but insight alone isn't enough. The true differentiator in an AI-driven world isn't just **what you know**, but **how fast you apply it**.

AI isn't just changing what knowledge is valuable; it's changing the **speed at which knowledge needs to be turned into action**. Those who wait for certainty will find themselves **perpetually behind**, while those who experiment, iterate, and execute rapidly will stay ahead.

In the next chapter, we move from **insight to execution**—from recognizing patterns AI can't see to **turning those patterns into career-defining action**.

THREE THINGS FOR **THIS** WEEK

Reality Check: You Can't Outsource Thinking to AI

1. Conduct a **'Decision Audit'**—how many of your daily decisions are **intuitive vs. data-driven?**
2. Identify one **AI-driven insight** that could improve your work—test it.
3. Start **thinking like an AI conductor**—where does your expertise add value to machine outputs?

10

AI-POWERED NETWORKING
SMARTER, NOT JUST BIGGER

*AI will transform reputation and relationships—
here's how to stay ahead.*

Professional networking is undergoing its most radical transformation in history—and most people are getting left behind. If AI is your collaborator, your next challenge isn't just using it—it's controlling how AI defines your value. Your résumé won't be read by a human; it will be scanned, ranked, and filtered by AI before anyone ever sees it. Your résumé won't be read by a human; it will be scanned, ranked, and filtered by AI before anyone ever sees it. AI has redefined professional networking. It's no longer about collecting contacts—it's about crafting the right signals so opportunity finds you.

The professionals who rely on outdated tactics—blindly attending events, amassing LinkedIn connections, and sending cold outreach messages—will soon find themselves **invisible** in an AI-mediated world. Meanwhile, those who **harness AI to amplify their social capital** will dominate the new professional landscape.

WHY NETWORKS MATTER MORE THAN EVER

The old model of professional success relied on credentials, tenure, and hierarchical advancement. But in an AI era, **your social capital—the resources embedded within your networks—has become your most valuable professional asset** (Lin, 2001). While machine intelligence handles increasingly complex technical tasks, the uniquely human connection becomes exponentially more valuable.

Traditional networking has always had limitations—geographic constraints, cognitive biases, and inefficient discovery. **AI doesn't replace networking—it supercharges it** by optimizing both strong and weak ties, fostering exploration across disconnected domains, and dramatically increasing access to innovation and opportunity.

AI flips the entire process on its head. Instead of casting a wide net and hoping for the best, professionals using AI are **engineering smarter, more targeted, and more meaningful connections.**

HOW AI TRANSFORMS PROFESSIONAL NETWORKING

Lisa, a management consultant, used to spend hours each month on what she called "spray and pray" networking—attending every industry mixer, sending generic follow-ups, and collecting connections like trading cards. Despite an impressive 5,000+

LinkedIn network, she struggled to translate this apparent network strength into tangible professional opportunities.

"I was doing everything the networking experts recommended," Lisa recalled. "Posting consistently, engaging with others' content, attending events, sending follow-ups. But the quality of relationships wasn't matching the effort I was investing."

Lisa didn't realize she was building what social capital theorists call "bridging capital"—many weak ties with little depth (Putnam, 2000). Her network missed the close trust of "bonding capital" and the powerful links of "linking capital" to decision-makers (Woolcock, 2001).

AI sharpened her approach with smarter targeting. Instead of hitting every event, she used AI tools to pinpoint professionals matching her skills, studying their work to spark real conversations, not just chit-chat (Microsoft & LinkedIn, 2024).

AI gives Lisa a powerful edge by pulling deep insights at scale. Before contacting someone, her AI tools dig up their latest projects, industry trends, and possible overlaps with her skills, letting her kick off with real talk instead of fluff (Microsoft & LinkedIn, 2024). This shift fits a trend where 75% of professionals use AI to work smarter, with many rethinking how it shapes their careers (Microsoft & LinkedIn, 2024).

Lisa's outreach gets personal with AI's help. She writes messages tied to each person's work, offering fresh takes instead of canned lines. Studies show tailored messages can lift response rates by 15% or more over generic ones (McKinsey & Company, 2023).

Most importantly, AI keeps Lisa's network humming without the hassle. It tracks follow-ups, recalls key details, and spots chances to reconnect, making her relationships strong and stress-free

(McKinsey & Company, 2023). This smart system could cut networking time by up to 30% across industries, experts say.

This isn't just about **more** connections. It's about **better** connections—building relationships **based on substance rather than chance.** The professionals who master this AI-powered approach **won't just grow their networks**—they'll create **real opportunities, faster and with greater impact than ever before.**

This transformation creates both threat and opportunity. The threat: traditional networking approaches will rapidly become obsolete as AI enables more targeted, personalized, and efficient relationship building. The opportunity: professionals who develop deliberate AI networking strategies will build more valuable social capital with less effort than ever before.

Lisa's shift wasn't just a personal breakthrough—it reflects a larger shift in how AI is reshaping social capital itself. To understand where professional opportunity is headed, we need to look at how AI is rewiring the very fabric of professional relationships.

SOCIAL CAPITAL IN THE AI AGE— WHAT'S CHANGED?

The Evolution of Professional Networks

For decades, researchers have recognized that social capital—the resources embedded in our social networks—drives professional success far more than individual ability alone (Lin, 2001). The classic model identified three critical forms:

- **Bonding capital**: Strong professional ties with close colleagues, mentors, and domain specialists who provide deep support and knowledge transfer
- **Bridging capital**: Weak ties with looser connections across different domains that foster innovative thinking and new opportunities
- **Linking capital**: Vertical connections to power structures, leadership circles, and decision-makers with disproportionate influence

In traditional networking, developing these different capital forms was a slow, fragmented process constrained by geography, cognitive capacity, and luck. The business card exchanges at industry events, reconnecting with former colleagues, and requesting introductions to key figures all operated through painfully manual processes with high failure rates.

THE AI-ENABLED EXPANSION OF SOCIAL CAPITAL

AI is fundamentally restructuring how social capital forms and functions by optimizing access, personalization, and discovery:

AI-enhanced bonding capital: Machine learning now helps curate deeper, more meaningful professional relationships by identifying exactly when and how to strengthen ties. David, a strategy consultant, used AI tools to recognize precisely which implementation specialists would complement his expertise, creating bonds that traditional networking would rarely uncover.

"The recommendations weren't just based on industry similarities," David explained. "The AI identified professionals whose specific methodological approaches aligned with mine in non-obvious ways, suggesting potential collaborations that I wouldn't have spotted through my own analysis."

AI-driven bridging capital: While Mark Granovetter (1973) first identified the "strength of weak ties" in driving innovation and opportunity, AI can now systematically identify and leverage these connections at scale. Instead of relying on serendipitous encounters, professionals can deploy algorithms that continually scan for high-potential weak-tie connections.

AI-enabled linking capital: Perhaps most transformatively, AI is democratizing access to power structures, funding sources, and leadership circles. Systems can now identify pathways to decision-makers that would have remained invisible in traditional networking approaches.

"I spent three years trying to connect with senior executives in my industry through traditional means," Lisa noted. "AI helped me identify precisely which thought leadership to share that would resonate with specific leaders, creating natural conversation opportunities that bypassed gatekeepers entirely."

AI AS AN EQUALIZER AND MULTIPLIER

AI isn't just shifting power to those who use it—it's redistributing access. For the first time, professionals outside of traditional power structures can strategically insert themselves into key networks, leveraging AI to amplify their reach and credibility. This democratization effect creates unprecedented opportunities for those previously excluded from elite networks due to geography, demographics, or institutional barriers.

A junior consultant in Singapore can now identify and meaningfully engage with industry leaders in New York. A self-taught developer can analyze and participate in conversations previously limited to Ivy League graduates. The multiplier effect of AI doesn't just make good networkers better—it can help level a playing field previously tilted toward those with inherited advantage.

This AI-powered expansion of social capital explains why the professionals embracing these tools experience such dramatic improvements in opportunity flow compared to those relying on traditional networking approaches.

The Shift from Static to Dynamic Networks

Traditional networking was fundamentally transactional and episodic. You'd attend an event, exchange information, follow up afterward, and then repeat the process elsewhere. Your professional identity remained relatively fixed, and relationship development followed predictable patterns.

AI-powered networking is transforming this model into a continuous, dynamic, and contextual process. Your professional identity is now constantly evolving based on your digital footprint, and relationship opportunities emerge in real-time rather than through planned engagements.

This shift aligns with what complexity leadership theorists describe as the move from hierarchical to adaptive organizational structures (Uhl-Bien, 2007). In complex adaptive systems, connections form and evolve based on emergent patterns rather than predetermined structures. AI accelerates this evolution by continually identifying and enabling these emergent patterns in professional networks.

THE POWER OF WEAK TIES AND STRUCTURAL HOLES IN AN AI WORLD

Why Weak Ties Drive Innovation

In his groundbreaking research, Mark Granovetter (1973) discovered that most breakthrough career opportunities come not from close professional contacts but from "weak ties"—loose connections outside your immediate circle. These weak ties provide access to novel information, perspectives, and opportunities that strong ties simply cannot.

Mark Granovetter's weak ties and Ronald Burt's (2004) structural holes theories proved that competitive advantage comes from bridging disconnected networks. AI now turns this from theory into strategy—systematically identifying valuable gaps in your network that would otherwise remain invisible.

AI dramatically enhances our ability to identify and leverage both weak ties and structural holes. Rather than leaving these valuable connections to chance, professionals can now systematically discover and cultivate them. AI can systematically identify gaps in networks where professionals can position themselves for outsized influence, turning Burt's theoretical advantage into a practical strategy.

David experienced this power directly when an AI matching algorithm identified a potential collaboration with an implementation specialist whose approach complemented his strategic methodology. "The algorithm connected me with a technical implementation specialist whose expertise perfectly complemented my strategic approach," David recalled. "I wouldn't have found her through my traditional network, as we had no connections in common and operated in different professional circles."

That single AI-mediated weak tie led to a collaboration that increased David's consulting revenue by 175% within eight months—a dramatic illustration of how AI can identify high-value bridging opportunities that traditional networking would likely miss.

AI AS AN INNOVATION NETWORK BUILDER

Beyond finding individual connections, AI excels at identifying patterns of innovation potential across networks. Michael Arena's (2018) adaptive space model describes how innovation flows between "operational networks" (focused on execution) and "entrepreneurial networks" (focused on exploration). The most innovative organizations create adaptive spaces where ideas can move freely between these different network types.

AI accelerates this movement by identifying potential connections between operational experts and entrepreneurial thinkers who might never otherwise connect. It can facilitate idea flow between efficiency-focused operational networks and innovation-driven entrepreneurial networks, creating bridges that spark new value. This explains why professionals leveraging AI often report not just more connections but more innovative collaborations that generate disproportionate value.

"What surprised me wasn't just finding more implementation partners," David noted. "It was discovering implementation specialists who approached problems from angles I'd never considered, sparking entirely new methodological innovations that neither of us would have developed independently."

THE NEW NETWORK ECONOMY: SIGNAL OVER NOISE

In the AI-powered network economy, the fundamental currency isn't connection count but signal strength—your ability to cut through information noise with distinctive expertise, perspective, or value creation.

David, the strategy consultant, discovered this power dynamic shift after analyzing his networking approach. Despite having fewer LinkedIn connections than many peers, he consistently developed more valuable professional relationships and opportunities by focusing on one thing: maximizing signal clarity for the right people rather than maximizing reach to all people.

"I noticed that every time I shared generic content—even when it reached a lot of people—it generated minimal meaningful engagement," David explained. "But when I shared specific, contextual insights related to my expertise, even if fewer people saw it, the right people responded with genuine interest."

This shift from broad reach to precision relevance is already visible in emerging patterns across professional platforms, with 75% of knowledge workers now using AI at work, including tools that enhance candidate assessment through digital insights (Microsoft & LinkedIn, 2024).

The second major shift is from generic content to contextual value. Generic "thought leadership" is being rapidly devalued as AI makes it easy to produce. Distinctive insights with specific contextual application create disproportionate engagement.

David observed this in his own content strategy: "When I used to write general articles about industry trends, they'd get a decent number of likes but almost no meaningful engagement. When I started sharing specific methodological approaches or distinctive perspectives on narrower topics, I'd get fewer total reactions but far more substantive responses from exactly the right people."

The third transformation is from reactive opportunity to strategic curation. Relationship development is shifting from opportunistic connection to deliberate curation of network composition aligned with specific professional objectives.

Finally, we're seeing a shift from human-only to AI-augmented engagement. Relationship maintenance is evolving from purely human memory and attention to systematically augmented engagement leveraging AI for context retention and opportunity identification.

The professionals who thrive in this new environment won't be those with the largest followings or the most connections. They'll be those who generate the strongest signal-to-noise ratio in their specific domains of expertise.

This transformation creates both dangers and opportunities. The danger: your existing network and influence approach may rapidly lose value as the rules of engagement change. The opportunity: you can build more meaningful, valuable relationships with significantly less effort by leveraging AI strategically.

CASE STUDY: THE NETWORKER WHO CUT THROUGH THE NOISE

Lisa had built what looked like a successful professional network—5,000+ LinkedIn connections, regular speaking engagements at industry events, and an active presence across digital platforms. But she struggled to translate this apparent network strength into tangible professional opportunities.

"I was doing everything the networking experts recommended," Lisa recalled. "Posting consistently, engaging with others' content, attending events, sending follow-ups. But the quality of relationships wasn't matching the effort I was investing."

At industry conferences, Lisa would collect dozens of business cards and dutifully add each person to her contact list. She'd send generic follow-up emails—"Great meeting you at the conference!"—but rarely developed these connections into meaningful relationships. Her content strategy focused on quantity and consistency rather than distinctive value or specific targeting.

Lisa's transformation began when she stopped focusing on network size and started developing a deliberate AI-augmented signal strategy:

1. **She used AI to analyze her existing content and engagement.** This revealed which topics, perspectives, and formats generated the strongest response from the specific professionals most relevant to her objectives.

"I discovered that certain aspects of my expertise—particularly around implementation methodology—generated disproportionate engagement from senior decision-makers," Lisa noted. "But these topics represented only about 15% of my content. I was diluting my signal with general industry commentary that didn't differentiate my perspective."

2. **She developed a domain-specific value map.** Rather than creating generic content, she identified specific knowledge gaps and perspective needs in her field, then systematically addressed these with distinctive insights.

3. **She built targeted engagement workflows.** Instead of scattering attention across her entire network, she used AI to identify the specific professionals most aligned with her interests, then developed personalized engagement approaches for each.

"I went from generic engagement—liking posts, leaving simple comments—to substantive contribution," Lisa explained. "I'd analyze someone's recent content, identify connection points to my expertise, and offer genuine perspective that advanced their thinking. The response difference was immediate and dramatic."

4. **She created an AI-augmented relationship maintenance system.** This preserved context across interactions, identified engagement opportunities, and ensured consistent follow-through without consuming excessive time.

The results were transformative. While her total network size remained relatively stable, the professional opportunities emerging from these relationships increased dramatically:

- Consulting inquiries from ideal clients increased by 340%
- Speaking invitations became more prestigious and better aligned with her expertise

- Collaboration opportunities with complementary professionals expanded significantly
- The time required for effective network maintenance decreased by approximately 60%

"The irony is that I'm spending less time on networking but creating far more valuable relationships," Lisa reflected. "By focusing on signal quality rather than network quantity, I've completely transformed my professional ecosystem."

Lisa's experience illustrates the core insight of this chapter: In the AI era, networking success comes not from connecting more but from connecting better—creating stronger signals for the right people while systematically filtering relationship noise.

THE AI NETWORKING PLAYBOOK

Building Your Network Intelligence System

The foundation of effective AI networking isn't any particular platform or tool. It's a comprehensive Network Intelligence System that leverages AI to enhance your relationship-building capabilities across all three forms of social capital—bonding, bridging, and linking.

This system has five integrated components:

1. **Network Mapping**: Creating visibility into your existing and potential professional ecosystem
2. **Value Matching**: Identifying specific value exchanges with the right connections
3. **Personalized Engagement**: Scaling authentic personalization across relationships

4. **Context Preservation**: Maintaining relationship depth without cognitive overload
5. **Opportunity Surfacing**: Systematically identifying collaborative possibilities

Let's explore how to build each component:

1. Network Mapping: Creating Visibility Into Your Ecosystem

Traditional networking operates with limited visibility—you know who you know, but have minimal insight into relationship patterns, strength distributions, or critical gaps.

David, a strategy consultant who struggled with networking as an introvert, began his transformation by applying analytical approaches to his professional connections. He exported his LinkedIn data and used network analysis tools to visualize his relationship ecosystem, revealing surprising patterns. "I discovered I had strong clusters in strategy development but almost no connections to implementation specialists," he noted. This visualization highlighted a critical gap between his strategic expertise and the technical experts who could execute his recommendations.

"The mapping showed me that 85% of my network consisted of other strategists or clients," David noted. "I had almost no connections to the implementation experts who could actually execute on my recommendations. This explained why my client engagements often stalled at the strategy phase."

This network gap represents what Burt (1992) would call a "structural hole"—a disconnection between network clusters that creates opportunity for those who bridge it. Strategic analysis enabled David to not only identify this structural hole but also develop a plan to bridge it.

AI-powered network mapping transforms this limitation, creating comprehensive visibility into your professional ecosystem:

Relationship Visualization: Use network visualization tools and LinkedIn's connection insights to map your existing network, uncovering clusters of relationships, strong versus weak ties, industry coverage, key gaps, and patterns of mutual engagement. Research from Harvard Business School shows that companies with well-connected employees—analyzed through LinkedIn data—are more innovative, suggesting that structured network analysis can reveal high-value opportunities for professionals too (Nagle et al., 2023).

Connection Opportunity Identification: Leverage AI-powered tools to identify high-value potential connections based on complementary expertise or resources, shared interests but different perspectives, strategic positions within relevant ecosystems, and potential for mutual value creation.

Network Health Assessment: Develop metrics for evaluating the strength and alignment of your network, including relevance to your current professional objectives, diversity across dimensions that matter for your field, reciprocity balance in value exchange, and growth in high-priority relationship categories.

Network mapping transforms relationship building from intuitive guesswork to strategic architecture. It allows you to see patterns and opportunities that remain invisible without systematic analysis.

2. Value Matching: Identifying Precision Exchange Opportunities

The core of meaningful networking isn't connection itself—it's value exchange. AI transforms this process by enabling precise identification of potential value flows between you and specific connections.

David developed a detailed inventory of his specific expertise, perspective, and access. Using AI tools, he identified potential connections where his strategic thinking would provide distinctive value to their execution challenges, creating natural reciprocity.

"I realized I had a unique perspective on translating complex strategy into implementable components," David explained. "This was precisely what many implementation specialists needed—creating a natural value exchange opportunity."

This approach aligns with social exchange theory, which suggests that relationships thrive when both parties perceive fair value exchange (Homans, 1958). The innovation is using AI to identify these potential value exchanges with much greater precision than human analysis alone could achieve.

AI enables three key aspects of value matching:

Value Inventory Development: Use AI to help create a comprehensive inventory of the value you can offer, including explicit knowledge and expertise you possess, tacit knowledge from your unique experience, access to resources or opportunities through your position, and perspective from your distinctive professional journey. Tools like Clarity.fm can help structure this inventory based on patterns in your expertise and experience.

Connection Value Analysis: Leverage AI to identify specific value potential connections might desire, based on their expressed professional interests and challenges, projects they're currently pursuing, questions they're exploring in their content, and gaps in their current professional ecosystem.

Precision Value Matching: Develop systems for identifying high-precision value exchanges, such as specific expertise you have that addresses their current challenges, unique perspectives you offer on topics they're exploring, resources you can access that complement their objectives, and connections you can facilitate aligned with their goals.

Generic value propositions like 'thought leadership' or 'industry expertise' are rapidly losing effectiveness. AI enables the identification of much more specific, contextual value exchanges that create truly meaningful connection opportunities.

3. Personalized Engagement: Authentic Connection at Scale

Traditional networking creates a painful tradeoff between personalization and scale. Deep personalization requires substantial time investment, while reaching more people typically requires sacrificing depth.

AI transforms this equation by enabling authentic personalization at scale—the ability to create genuinely customized engagement without proportional time investment.

David employed CrystalKnows to analyze the communication preferences of potential connections before outreach. "The tool predicted that a particular implementation director preferred direct, data-driven communication with minimal small talk," David explained. "I crafted my outreach accordingly, focusing immediately on the specific value exchange rather than starting with pleasantries."

This personalized approach led to an immediate response and ultimately a collaboration that secured a $500,000 contract—an opportunity David would have likely missed with his previous generic networking approach. The specific personality insights provided by the AI tool enabled him to tailor his communication in ways that resonated with the director's preferences, creating an immediate rapport that bypassed the usual networking discomfort.

AI enables personalized engagement through:

Research Augmentation: Use AI to develop rich contextual understanding before engagement, including a comprehensive analysis of their professional history and focus, identification of their core interests and perspective, recognition of their current projects and challenges, and understanding of their communication preferences and style.

Personalization Framework: Develop systematic approaches to meaningful customization by identifying specific aspects of their work that connect to yours, finding genuine points of intellectual or professional curiosity, recognizing opportunities for authentic appreciation, and crafting engagement that demonstrates actual understanding.

Outreach Enhancement: Create workflows that maintain a genuine voice while leveraging AI, including draft generation based on your authentic communication style, refinement processes that ensure your distinctive perspective, customization approaches that preserve relationship authenticity, and review systems that maintain quality at increased scale. Tools like Unito or Zapier can create automated workflows that gather relevant information about potential connections before outreach.

The goal isn't automating personalization—it's amplifying your genuine relationship-building capacity. Effective AI networking doesn't replace your authentic voice; it expands how many meaningful connections you can develop with that voice.

4. Context Preservation: Maintaining Relationship Depth

One big limit in traditional networking is human memory—it's tough to keep track of details across many relationships over time.

David's approach highlights a growing shift in how professionals build relationships, with AI playing a bigger role in connecting people. Tools like Clay.com help him manage context and match with the right contacts, moving beyond random meetings or location-based ties.

Research from McKinsey shows AI is already boosting work efficiency by up to 30%, pointing to a future where algorithms increasingly guide professional networking (McKinsey & Company, 2023).

AI makes this possible through:

- Interaction Memory: Systems that log and sort past conversations, interests, commitments, and personal details for deeper ties.
- Engagement Timing: Plans for when to reconnect, based on relationship history, key moments, and a balanced network approach to keep ties alive.
- Value Continuity: Ways to keep value flowing by tracking what you've offered, spotting new opportunities as they

grow, and building trust over time without it feeling forced.

- This turns networking into an ongoing process, not just one-off chats, letting you nurture more genuine relationships than memory could handle alone.

5. Opportunity Surfacing: Systematic Collaboration Identification

Perhaps the most powerful aspect of AI networking is systematic opportunity identification—the ability to spot potential collaboration, learning, or value-creation possibilities across your network.

Traditional networking discovers opportunities primarily through chance—the right conversation at the right time. AI networking transforms this into a systematic process.

David used Lunchclub's AI matching algorithm to identify specific collaboration opportunities within and beyond his network. The system regularly suggested precision-matched connections based on complementary expertise and mutual interests, eliminating the randomness of traditional networking.

"One match in particular transformed my business," David recalled. "The algorithm connected me with a technical implementation specialist whose expertise perfectly complemented my strategic approach. I wouldn't have found her through my traditional network, as we had no connections in common and operated in different professional circles."

This collaboration led to the development of a unique strategic implementation methodology that became David's signature offering, increasing his consulting revenue by 175% within eight months. The precision of the AI matching created a partnership that would have been unlikely to form through traditional networking approaches.

This approach aligns with Arena's (2018) adaptive space model by systematically connecting operational and entrepreneurial networks, fostering innovation that neither group would develop independently.

AI enables opportunity surfacing through:

Complementarity Analysis: Develop systems for identifying potential synergies, including complementary expertise that could combine to address challenges, shared interests approached from different perspectives, resource access that creates mutual advantage when combined, and positioning that enables unique collaborative possibilities.

Trigger Event Monitoring: Create frameworks for recognizing opportunity moments such as professional transitions that change relationship possibilities, project initiations that create collaboration potential, question exploration that aligns with your expertise, and challenge emergence where you could provide value.

Introduction Orchestration: Build approaches for facilitating valuable connections by identifying specific value potential in connection between others, creating contextualized introduction frameworks, ensuring mutual benefit in facilitated relationships, and monitoring engagement to refine future connection decisions.

"Systematic opportunity surfacing transforms networking from passive collection to active curation," observes network strategist

Dr. James Park. "It allows you to continuously create value through your network rather than simply maintaining connections."

This transforms your network from a static asset to a dynamic value creation system that continuously surfaces new possibilities.

CASE STUDY: THE COMPLETE NETWORK INTELLIGENCE SYSTEM

David, a strategy consultant, had struggled with networking throughout his career. As an introvert who disliked superficial conversation, traditional networking approaches felt inauthentic and draining. Yet he recognized that high-quality relationships were essential for professional success.

David's approach to networking changed when he stopped trying to 'network better' and instead built a Network Intelligence System—a methodical, AI-powered approach that played to his strengths rather than forcing him into uncomfortable networking situations. David's approach was methodical and tool-driven:

1. Network Mapping: David used NodeXL to analyze his existing LinkedIn connections, revealing surprising patterns. "I discovered I had strong clusters in strategy development but almost no connections to implementation specialists," he noted. This visualization highlighted a critical gap between his strategic expertise and the technical experts who could execute his recommendations.

2. Personality Insights: He employed CrystalKnows to analyze the communication preferences of potential connections before outreach. "The tool predicted that a particular implementation director preferred direct, data-driven communication with minimal small talk," David explained. "I crafted my outreach accordingly, focusing immediately on the specific value exchange rather than starting with pleasantries."

This personalized approach led to an immediate response and ultimately a collaboration that secured a $500,000 contract—an opportunity David would have likely missed with his previous generic networking approach. The specific personality insights provided by the AI tool enabled him to tailor his communication in ways that resonated with the director's preferences, creating immediate rapport that bypassed the usual networking discomfort.

3. Content Analysis: Using Postwise's AI content analyzer, David examined which types of posts generated the most meaningful engagement from his target audience. "I discovered that my detailed breakdowns of implementation roadblocks received 320% more engagement from technical experts than my general strategy content," he explained. This insight allowed him to focus his limited content creation energy on topics that actually resonated with the connections he needed most.

4. Relationship Management: David implemented Clay.com to maintain contextual understanding across his key relationships. Before any follow-up conversation, he could review the entire history of his interactions with that person, including topics discussed, interests mentioned, and previous value exchanges. "I no longer had to rely on my fallible memory to maintain relationship continuity," David noted. "The system ensured I could pick up conversations naturally months later, creating a sense of thoughtfulness that strengthened connections."

5. Opportunity Surfacing: Perhaps most importantly, David used Lunchclub's AI matching algorithm to identify specific collaboration opportunities within and beyond his network. The system regularly suggested precision-matched connections based on complementary expertise and mutual interests, eliminating the randomness of traditional networking.

"One match in particular transformed my business," David recalled. "The algorithm connected me with a technical implementation specialist whose expertise perfectly complemented my strategic approach. I wouldn't have found her through my traditional network, as we had no connections in common and operated in different professional circles."

This collaboration led to the development of a unique strategic implementation methodology that became David's signature offering, increasing his consulting revenue by 175% within eight months. The precision of the AI matching created a partnership that would have been unlikely to form through traditional networking approaches.

The results transformed David's professional trajectory. Within one year of implementing his Network Intelligence System:

- His outreach response rates increased from under 20% to consistently above 85%
- He established collaborative relationships with exactly the implementation specialists his practice needed
- His LinkedIn content engagement increased by 340% while actually posting less frequently
- He facilitated 28 high-value connections that strengthened his position as a trusted resource
- His consulting revenue increased by 175% through new collaborative opportunities

- He reduced his networking time investment by approximately 60% while generating significantly better results

"The system hasn't made me someone I'm not—I'm still an introvert who values depth over breadth," David reflected. "What it's done is amplify my authentic relationship-building approach, allowing me to connect meaningfully with exactly the right people without the exhaustion of traditional networking."

David's experience illustrates the power of a complete Network Intelligence System. By systematically addressing each component, he transformed networking from a dreaded obligation to a structured process for building valuable relationships aligned with his natural style and professional objectives.

THE AUGMENTED NETWORKER—THRIVING IN THE AI ERA

The Three Rules of AI-Augmented Networking

As AI continues transforming professional networking and reputation, success requires following three fundamental rules:

1. **Be Findable**: Contribute valuable insights publicly (thought leadership, open-source projects, content) that make your expertise discoverable by both humans and AI systems.
2. **Be Discoverable**: Leverage AI-driven platforms to amplify your expertise and ensure you appear in the right searches and recommendations.

3. **Be Adaptable**: Constantly expand network frontiers, blending bonding, bridging, and linking capital to create a dynamic professional ecosystem.

The professionals who master these rules won't just have more connections—they'll develop the precise relationships that drive opportunity in an AI-mediated world.

ACTIONABLE STEPS FOR BUILDING AI-AUGMENTED SOCIAL CAPITAL

Here are five immediate actions to supercharge your networking with AI:

1. **Map Your Current Network Reality**: Use AI tools like NodeXL or LinkedIn's connection insights to visualize your professional web. Spot clusters (e.g., industry groups), gaps (e.g., missing decision-makers), and opportunities tied to your goals. Research shows companies with well-mapped employee networks—via LinkedIn data—boost innovation, hinting at similar gains for individuals (Nagle et al., 2023).

2. **Develop Your Value Exchange Inventory**: List what you bring to the table and what you need from others. Tools like Clarity.fm can analyze your expertise to sharpen this list. Aim for precise matches—think targeted value over broad networking—to spark meaningful ties.

3. **Build Your Personalization Workflows**: Set up AI-driven systems for authentic outreach at scale. Tools like Unito or Zapier can pull data on potential connections—think recent projects or posts—before you reach out. McKinsey notes personalized AI tools lift efficiency by 15-30%,

suggesting better response rates too (McKinsey & Company, 2023).

4. **Implement Basic Context Preservation**: Use tools like Clay or HubSpot to track interactions and keep relationships warm. Log talks, interests, and follow-ups for seamless continuity. McKinsey's data shows AI can cut work time by up to 30%, letting you nurture more ties without the mental load (McKinsey & Company, 2023).

5. **Conduct a Digital Reputation Audit**: Check how you're seen online with tools like Brand24 or Mention. Compare your real skills to your digital footprint and fix gaps. A strong online presence draws opportunities—75% of professionals already use AI to stand out, per recent trends (Microsoft & LinkedIn, 2024).

These foundational steps will immediately strengthen your professional relationships while preparing you for more advanced strategies as AI capabilities and your own strategic thinking evolve.

THE NEW CURRENCY OF SUCCESS

Your social network has become your career moat—and AI makes it wider, deeper, and more strategically valuable. In a world where technical skills are increasingly automatable, your unique combination of relationships, perspective, and expertise creates competitive advantage that AI can enhance but never replace.

The best professionals don't just build large networks—they cultivate the right balance of weak and strong ties. Weak ties expose you to new ideas, industries, and opportunities you wouldn't find in your immediate circles. Strong ties, however, provide the trust and resources to execute. The key is managing both—seeking out novelty while ensuring you have the relationships to bring those ideas to life.

This distinction between exploratory and execution-oriented networking is becoming even more critical as the job landscape undergoes transformation. According to *The Future of Jobs Report 2025*, AI and automation will disrupt nearly **44% of core workplace skills** by 2028, significantly altering the balance between technical execution and human-centric collaboration (World Economic Forum, 2024). As AI automates increasingly complex technical tasks, professionals who can **blend weak-tie discovery (novelty) with strong-tie execution (implementation) will have an outsized advantage**. AI-optimized networks will increasingly drive **opportunity flow, career mobility, and professional visibility**, making adaptability in social capital a defining skill of the AI era.

THE AI-DRIVEN DUAL NETWORK STRATEGY

AI enables professionals to balance exploration (weak ties) and execution (strong ties) with unprecedented efficiency. Those who master this dual-network strategy will have the agility to innovate and the credibility to execute—outpacing competitors stuck in outdated networking models.

Take inventory of your current network—are you over-relying on strong ties that reinforce the status quo? Or do you have too many weak ties without the deep relationships needed to execute? AI can help you correct this imbalance and optimize your network for both discovery and action.

For the first time, professionals outside of traditional power structures can strategically insert themselves into key networks, leveraging AI to amplify their reach and credibility. A self-taught developer in Nairobi, a mid-career professional in a rural area, or a freelancer outside major corporate hubs can now build and leverage relationships that were previously accessible only to those within privileged circles.

AI isn't just reshaping how we network—it's changing the very structure of our networks. It allows us to discover and cultivate weak ties at an unprecedented scale, ensuring a constant flow of new ideas. Simultaneously, AI helps us maintain and deepen strong ties, ensuring we can execute those ideas faster and more effectively than ever before.

Tools like Lunchclub's algorithmic matching systematically identify and close structural holes in your network, connecting you with precisely the right people across disconnected clusters. These connections are no longer left to chance but can be strategically cultivated through AI-powered recommendations.

In an AI-driven world, career success isn't just about individual skills—it's about your position within networks that cultivate both novelty and execution. AI-augmented professionals don't just see opportunities before others—they execute on them faster, with greater precision, and at a scale that was previously impossible. This **execution velocity** is the real competitive advantage in the AI era.

The professionals who thrive in AI-driven networks aren't the loudest or the most connected—they're the ones who **signal value with precision**. Your network isn't just who you know—it's **who recognizes your expertise when it matters most**.

In the AI age, opportunity won't come from a cold email or a lucky break—it will come from how well AI recognizes your expertise before a human ever does.

Chapter 10: The New Leverage—AI, Social Capital, and the Future of Professional Networks

THREE THINGS FOR **THIS** WEEK

Reality Check: It's Not About Who You Know—It's About Who Knows Your Value

1. Identify **one AI-related trend** that will reshape networking in your field—learn about it.
2. Look at your industry's AI adoption—**are you ahead or behind?** What's your response?
3. Define one **actionable step** you can take this week to **future-proof** your network.

11

The Future of Work: The Only Constant is Change So You Must Be Too

There is no endpoint. Those who stay adaptable will win—here's how to be one of them.

"When does it end?" the executive asked me.

We had just finished a workshop on AI adaptation strategies at his company, and he looked exhausted. "First it was digital transformation, then remote work, now AI revolution. When do we reach the point where things stabilize and we can just execute?"

In the AI era, stability isn't just elusive—it's a mirage. The only certainty is constant reinvention, and professionals who master adaptation will outpace those who cling to outdated certainties.

I gave him a hard truth: "Never."

The executives who understand this will build companies that thrive. The ones who don't? They'll spend the next decade playing defense, watching their industries—and careers—get rewritten around them.

This isn't just about AI. It's about the death of stability itself.

THE ACCELERATION OF CHANGE

The uncomfortable truth about the future of work isn't just that things are changing—it's that the pace of change itself is accelerating. McKinsey reports AI could automate 30% of today's work tasks, driven by leaps in data and computing power (McKinsey & Company, 2023, p. 12). We're not experiencing a temporary disruption that will eventually settle into a new stable equilibrium. We're entering an era where constant adaptation is the only equilibrium.

This reality is creating a profound "knowing-doing gap" across organizations. Deloitte's 2023 survey of 2,840 leaders found that 73% see the need to align human skills with tech, yet only 9% are making headway (Deloitte Insights, 2023). The gap between awareness and action is the breeding ground for professional obsolescence.

This isn't a popular message. Humans are naturally wired to seek stability. We want to learn the rules, master the game, and then efficiently execute. The idea that the rules will continuously change—that mastery itself must be redefined as adaptive capacity rather than static expertise creates profound psychological and organizational discomfort.

But the professionals and organizations that thrive in the coming decades won't be those who resist this reality or hope for its reversal. They'll be those who develop what I call perpetual transformation capacity—the ability to continuously reinvent themselves in response to (and ideally ahead of) changing conditions.

Despite the rapid adoption of AI, its business impact remains far from guaranteed. Companies are projected to invest heavily in AI initiatives—PwC estimates up to $15.7 trillion by 2030 (PwC, 2021)—yet many struggle to translate that investment into measurable returns.

As Stanford's Erik Brynjolfsson points out, there's often a "Productivity J-Curve" with transformative technologies like AI, where initial investments may show limited returns before eventually scaling up to create significant value (Brynjolfsson et al., 2019). This underscores a critical shift: AI isn't a guaranteed business advantage; it's a tool that must be strategically integrated.

Lisa, a product manager at a technology company, embodies the adaptation approach. "I stopped thinking of my career as a progression toward a destination," she told me. "Instead, I see it as a continuous evolution with no fixed endpoint. Every six months, I reassess which of my capabilities are appreciating in value and which are depreciating, then adjust my development accordingly."

This perspective transformed Lisa's relationship with change from threat to opportunity. While colleagues resisted each new transformation initiative, she systematically explored how emerging technologies and practices could enhance her impact. When AI began transforming product management, she was already experimenting with its capabilities—not because she had predicted exactly how it would develop, but because she had built continuous adaptation into her professional operating system.

"The people who struggle most are those waiting for the 'new normal,'" Lisa observed. "They want to endure the current change, then settle into executing whatever comes next. But there is no 'after.' The only sustainable approach is getting comfortable with perpetual transformation."

This final chapter isn't about specific technologies or trends that will shape tomorrow's workplace. Such predictions inevitably fall short as innovation accelerates. Instead, it's about developing the meta-capability that transcends specific changes: the capacity to continuously reinvent yourself regardless of which specific disruptions emerge.

THE META-SKILL OF PERPETUAL TRANSFORMATION

Why Adaptation Velocity Is the Ultimate Competitive Advantage

Throughout this book, we've explored specific strategies for thriving amid AI disruption—from building skill stacks to developing AI sidekick strategies to reimagining professional networking. But these specific approaches, while powerful today, will themselves evolve as technologies and practices advance.

Adaptation velocity isn't just a competitive advantage—it's the dividing line between those who will lead the AI era and those who will be left behind.

Most professionals still believe the future will reward expertise. It won't. It will reward reinvention.

With skills fading in five years or less (Deloitte Insights, 2021), the ability to redefine your professional value—over and over again—is what separates thriving professionals from those stuck in jobs that no longer exist.

As work becomes increasingly "boundaryless"—where traditional boundaries around jobs, workplaces, and workforce models blur and dissolve—the emphasis shifts from credentials to capabilities. What's striking is that while most organizations recognize this shift, Deloitte finds only 13% prioritize adaptability over old metrics (Deloitte Insights, 2023)—creating a massive opportunity for professionals who can adapt faster than their organizations.

We're seeing a fundamental split in professional trajectories. It's not primarily about technical skills, credentials, or even intelligence in the traditional sense. It's about the speed at which people can absorb new realities and reconfigure their approach in response.

This adaptation velocity manifests across multiple dimensions:

1. **Learning Agility**: How quickly you can acquire and apply new knowledge as needs emerge
2. **Identity Flexibility**: How readily you can redefine your professional self-concept as conditions evolve
3. **Comfort with Ambiguity**: How effectively you can function amid uncertainty and incomplete information
4. **Recovery Resilience**: How rapidly you rebound from inevitable setbacks during transformation
5. **Opportunity Recognition**: How quickly you identify emerging possibilities others don't yet see

These capabilities aren't fixed traits but learnable skills. The professionals who thrive don't necessarily start with higher adaptation velocity. They deliberately develop it through specific practices and mindset shifts.

Why Most Professional Development Is Fundamentally Obsolete

Most professional development approaches are fundamentally misaligned with today's rapid pace of change. This misalignment stems from three critical flaws in conventional approaches:

1. **The Destination Fallacy**: Traditional development assumes a stable professional endpoint where skills, once acquired, retain their value indefinitely. In reality, the World Economic Forum estimates that 54% of workers need reskilling by 2025 (World Economic Forum, 2020).

2. **The Credential Trap**: Conventional approaches prioritize formalized learning and credentials over practical adaptation. Yet with the accelerating pace of technological disruption, by the time many formal credentials are earned, the underlying skill requirements have already evolved.

3. **The Expertise Paradox**: Traditional development seeks to build deep domain expertise, but increasing specialization can actually reduce adaptation velocity. Organizations are finding that professionals who can operate across multiple domains often adapt more effectively to disruption than narrowly specialized experts.

The fundamental problem isn't the quality of professional development—it's the underlying model. We're preparing professionals for a world of predictable, linear progression when they're actually entering an environment of continuous, non-linear transformation.

THE THREE MINDSETS OF PERPETUAL TRANSFORMATION

Adaptation doesn't just happen—it's cultivated intentionally. Through my interviews with hundreds of professionals thriving amid technological change, I've identified three fundamental mindsets that separate those who merely survive disruption from those who harness it as fuel for growth:

Mindset	Key Shift	Core Practice
Evolutionary	Best Practice → Next Practice	Horizon Scanning, Expertise Audits
Portfolio	Single Path → Multiple Options	Asymmetric Opportunity, Optionality Creation
Network	Solo Expertise → Collaborative Value	Knowledge Exchange, Connection Facilitation

1. The Evolutionary Mindset: Constant Experimentation

Traditional professional development follows a linear accumulation model: learn established best practices, implement them effectively, and gradually accumulate expertise. The evolutionary mindset takes a fundamentally different approach based on continuous experimentation:

From Best Practice to Next Practice Rather than focusing primarily on implementing established approaches, the evolutionary mindset continuously explores potential innovations through deliberate experimentation.

279

"I maintain at least three professional experiments at all times," explains marketing director Elena Rodriguez. "Some are small tweaks to existing approaches; others explore entirely new possibilities. This continuous experimentation means I'm never wholly dependent on current best practices, which inevitably become obsolete."

From Failure Avoidance to Learning Acceleration Instead of prioritizing error prevention, the evolutionary mindset treats setbacks as essential data for rapid adaptation.

"I've reframed how I think about professional 'failures,'" shares product developer Michael Chen. "They're not evidence of inadequacy but acceleration opportunities for my adaptation. The question isn't whether something will work perfectly, but what I'll learn that improves my next iteration."

From Expertise Defense to Knowledge Evolution Rather than protecting current knowledge investments, the evolutionary mindset continuously evaluates which capabilities to grow, maintain, or shed.

"Every quarter, I conduct a deliberate expertise audit," explains financial analyst Thomas Wu. "I identify which knowledge areas are appreciating versus depreciating in value, then adjust my learning investments accordingly. This prevents the common trap of overinvesting in maintaining declining expertise."

The evolutionary mindset transforms professional development from a linear progression to a continuous adaptation process. It trades the comfort of mastery for the advantage of perpetual evolution.

2. The Portfolio Mindset: Strategic Diversification

Traditional career planning focuses on optimizing a single professional path. The portfolio mindset takes a fundamentally different approach based on strategic diversification:

From Single Path to Multiple Options Rather than pursuing one optimal career trajectory, the portfolio mindset develops multiple potential paths that can be activated as conditions change.

"I've stopped asking 'What's my five-year plan?'" explains consultant Sarah Johnson. "Instead, I ask 'What portfolio of capabilities and relationships am I building, and what options does this portfolio create?' This shift has transformed how I make development decisions."

From Risk Minimization to Asymmetric Opportunity Instead of avoiding professional risk, the portfolio mindset seeks positive asymmetries—situations where the potential upside significantly exceeds the downside.

"I deliberately pursue projects with open-ended upside and limited downside," shares software developer David Rodriguez. "These asymmetric opportunities create disproportionate growth potential without existential professional risk."

From Income Maximization to Optionality Creation Rather than optimizing solely for current compensation, the portfolio mindset prioritizes creating future possibilities through strategic capability development.

"I sometimes take roles that pay less but develop capabilities I believe will appreciate dramatically in value," explains operations specialist Maria Chen. "This investment approach to professional choices creates exponentially more options over time than simply maximizing current income."

The portfolio mindset transforms career development from path optimization to strategic option creation. It trades the clarity of a single trajectory for the resilience of multiple potential paths.

3. The Network Mindset: Value Through Connection

Traditional professional value creation focuses on individual capability. The network mindset takes a fundamentally different approach based on connection facilitation:

From Solo Expertise to Collaborative Creation Rather than developing all necessary capabilities internally, the network mindset creates value by connecting complementary expertise.

"I've stopped trying to master everything myself," explains creative director James Park. "Instead, I focus on developing enough understanding across domains to facilitate effective collaboration between specialists. This connector role creates far more value than attempting comprehensive personal expertise."

From Knowledge Hoarding to Generous Exchange Instead of protecting expertise as competitive advantage, the network mindset creates advantage through active knowledge sharing.

"I systematically share what I'm learning and connect others to relevant resources," shares researcher Lisa Thompson. "This generosity paradoxically strengthens my position because it makes me a valuable node in knowledge exchange networks."

From Fixed Position to Dynamic Reconfiguration Rather than occupying a stable role, the network mindset continuously reconfigures professional positioning as collaborative needs evolve.

"I think of myself less as having a fixed professional identity and more as a flexible resource that reconfigures based on project needs," explains consultant Michael Torres. "This adaptability allows me to participate in a much wider range of opportunities than a static positioning would permit."

The network mindset transforms professional value from individual contribution to connection facilitation. It trades the security of self-sufficiency for the expanded impact of collaborative creation.

THE ADAPTATION GAP: WHY PROFESSIONALS FACE ACCELERATING OBSOLESCENCE

Picture this: a marketing executive sits in a board meeting as leaders discuss AI-powered customer personalization. She nods along, acknowledging AI's importance—then returns to her department and continues business as usual. Six months later, her team is frantically trying to catch up as competitors deploy AI solutions she could have been developing all along.

This is the adaptation gap in action—the growing distance between the pace of change and our capacity to respond to it.

As organizations accelerate their adoption of AI and related technologies, this gap continues to widen. McKinsey's 2023 report indicates that AI adoption is accelerating rapidly, with some organizations seeing efficiency gains of up to 30% in specific domains (McKinsey & Company, 2023). This "adaptation gap" creates both unprecedented risk and opportunity.

Deloitte (2023) identifies an alarming "imagination deficit"—AI is evolving faster than professionals' ability to envision new applications and possibilities. This deficit reflects a broader pattern: it's not that people don't see change coming, it's that they can't imagine how to transform themselves to capitalize on it.

The scale of this imagination deficit is striking. Deloitte's 2023 data shows that 73% see the need, but only 9% act (Deloitte Insights, 2023). This gap between awareness and implementation represents both a significant challenge and opportunity for forward-thinking professionals.

What's more, workers themselves recognize this gap. According to the same research, 76% of workers want help envisioning their future, but only 43% get it (Deloitte Insights, 2023). This discrepancy highlights the very real impact of the imagination deficit on the workforce and underscores the urgent need for both organizations and individuals to address it.

What separates high-velocity adapters from those vulnerable to obsolescence isn't raw intelligence or even specific technical skills—it's their approach to transformation itself.

CASE STUDY: THE PERPETUAL TRANSFORMER

Rachel's career illustrates the power of these three mindsets in creating extraordinary adaptation velocity.

As a marketing professional, Rachel had built a successful career following traditional advancement paths—developing expertise in established methodologies, pursuing linear promotion through organizational hierarchies, and cultivating a clear professional identity.

When digital transformation began disrupting marketing, many of her colleagues responded by doubling down on existing expertise or reluctantly learning specific new tools. Rachel took a fundamentally different approach based on perpetual transformation:

The Evolutionary Mindset in Action: Rather than viewing digital marketing as a specific skill set to acquire, Rachel established continuous experimentation as her core professional practice:

- She maintained a personal innovation lab where she tested emerging approaches before client application
- She reframed setbacks as essential data that accelerated her learning curve
- She conducted quarterly expertise audits, systematically evolving her capability portfolio

The Portfolio Mindset in Action: Instead of pursuing a single career path, Rachel developed a diversified professional strategy:

- She maintained multiple potential trajectories ranging from creative direction to analytics leadership to entrepreneurial ventures
- She deliberately sought asymmetric opportunities where experimental approaches could create a disproportionate impact
- She prioritized capability development over immediate compensation, building expertise in emerging areas before their value was widely recognized

The Network Mindset in Action: Rather than trying to master every aspect of rapidly evolving marketing technology, Rachel focused on connection value:

- She developed sufficient understanding across domains to facilitate effective collaboration between specialists
- She created systems for continuous knowledge sharing that positioned her as a valuable network node
- She adopted flexible professional positioning that allowed her to reconfigure her role based on project needs

The results were remarkable. While many skilled marketers struggled to maintain relevance amid continuous disruption, Rachel thrived through multiple technological transformations—from social media to mobile to programmatic to AI-powered marketing.

"The specific technologies and methodologies kept changing, but my approach to adaptation remained consistent," Rachel explained. "By focusing on developing transformation capacity rather than mastering any particular marketing approach, I built sustainable advantage regardless of which specific disruptions emerged."

Rachel's experience illustrates the central insight of this chapter: The most valuable professional meta-skill isn't expertise in any particular domain or technology. It's the capacity for perpetual transformation—the ability to continuously reinvent yourself as conditions evolve.

DESIGNING YOUR TRANSFORMATION OPERATING SYSTEM

The Five Practices of High-Velocity Adapters

Mindsets are the foundation, but systems make transformation sustainable. The professionals who consistently outpace disruption don't rely on willpower or inspiration—they build robust practices that make adaptation their default mode:

Think of these practices as your transformation operating system—the background processes that keep you evolving even when you're focused on today's deliverables.

1. Horizon Scanning: Systematic Future Monitoring

High-velocity adapters don't just respond to change after it arrives; they systematically monitor for early signals that allow proactive adaptation:

Multi-Horizon Monitoring

- **Horizon 1 (0-12 months)**: Imminent changes already beginning to impact your field
- **Horizon 2 (1-3 years)**: Emerging trends gaining momentum but not yet mainstream
- **Horizon 3 (3-7 years)**: Early signals that might represent future transformations

Cross-Domain Exploration

- Adjacent industries experiencing disruption that could spread to yours

- Academic research that might eventually translate to practical application
- Cultural or demographic shifts that could change stakeholder expectations

Pattern Recognition

- Recurring themes across different information sources
- Accelerating adoption trajectories that suggest gathering momentum
- Unexpected combinations that might create emergent possibilities

Effective horizon scanning isn't about predicting exactly what will happen. It's about expanding your awareness of potential developments so you can adapt more quickly when directional changes become clear.

Think of Carlos, a finance professional who maintained a simple but consistent scanning practice: every Friday, he spent 30 minutes reviewing developments across fintech, blockchain, and AI. When generative AI tools first appeared, he recognized their potential for financial modeling months before they hit the mainstream—giving him time to experiment and build competency ahead of the curve.

2. Learning Acceleration: Systematic Capability Development

High-velocity adapters don't just learn continuously; they develop systematic approaches to accelerate knowledge acquisition and application:

Meta-Learning Optimization

- Identify your specific learning preferences and adapt approaches accordingly
- Create standardized processes for rapidly acquiring new knowledge areas
- Build connection systems that integrate new learning with existing knowledge

Deliberate Knowledge Experimentation

- Develop quick implementation approaches for testing new concepts
- Create feedback systems that assess application effectiveness
- Establish refinement processes that evolve your understanding

Knowledge Network Leveraging

- Build connections with complementary expertise domains
- Create collaborative learning structures that multiply individual capacity
- Develop knowledge exchange systems that accelerate mutual development

The difference between average and exceptional learners isn't intelligence but approach. High-velocity adapters develop systematic processes that dramatically accelerate knowledge acquisition and application.

Maya, a project manager, exemplifies this approach. When her organization adopted an AI-powered workflow platform, she didn't just read the manual—she created a 30-day micro-learning experiment. Each day, she tested one new feature and documented its potential applications. By the time her colleagues were scheduling training sessions, Maya was already leveraging the platform to automate routine tasks and identify process improvements that saved her team 15 hours weekly.

THE PERPETUAL TRANSFORMER'S MANIFESTO

As industrial sectors face ongoing disruption driven by AI and related technologies, professionals face both unprecedented challenges and extraordinary opportunities.

The future belongs not to those who predict specific changes most accurately or respond to particular disruptions most effectively. It belongs to those who develop the meta-capability of perpetual transformation—the capacity to continuously reinvent themselves regardless of which specific changes emerge.

This perpetual transformation capacity creates three distinct advantages:

1. **Resilience Advantage**: The ability to maintain effectiveness amid disruption that diminishes others' performance
2. **Timing Advantage**: The capacity to identify and act on emerging opportunities before they become obvious to others
3. **Evolution Advantage**: The capability to continuously develop in ways that create expanding rather than diminishing returns

What we're really talking about is the ability to thrive through change rather than despite it. The most adaptable professionals don't just endure transformation—they leverage it as a catalyst for unprecedented growth.

THE AI-PROOF PROFESSIONAL

AI is changing not just how we work, but how we connect, build influence, and create value. The professionals who thrive in this new landscape won't be those who passively wait for opportunities—they'll be the ones who intentionally design their careers, using AI as an amplifier, not a substitute.

Your network is no longer just about who you know—it's about how you create and demonstrate value in a digital-first world. The strongest careers in the AI era won't be built on automation alone, but on human relationships, adaptability, and a strategic understanding of AI's role as a force multiplier.

But here's the truth: technology will keep evolving—faster than you can predict. With many skills now having a half-life of just 5 years or less (Deloitte Insights, 2021), continuous adaptation isn't optional—it's essential.

The question isn't whether AI will disrupt your field—it already is. The question is whether you'll let it define your career, or whether you'll take control.

The ultimate competitive advantage in the AI era isn't just about who you know. It's not about what you've done in the past. It's about your ability to evolve.

THREE CHOICES DEFINE YOUR FUTURE:

Reposition Your Value

Are you operating at a level where AI complements you rather than replaces you?

Amplify with AI

Are you leveraging AI as a tool to multiply your impact, or are you waiting to be disrupted?

Execute for Impact

Are you taking action today, testing, experimenting, and learning—or are you waiting for certainty that will never come?

AI isn't coming for your job. Complacency is. If you don't define your career, someone—or something—else will. But here's the good news: You've made it through this book. You're already ahead of the curve. Now, the challenge isn't survival—it's growth. The future belongs to those who embrace the shift, harness the change, and step forward with intention.

The professionals who win are not just adapters. They are architects. In the next chapter, you'll meet one.

Transformation Operating System: A Guide to Professional Adaptation in the AI Era

The Three Mindsets of Perpetual Transformation

Mindset	Key Shift	Core Practices	Success Indicators
Evolutionary	Best Practice → Next Practice	• Horizon scanning • Expertise audits • Deliberate experimentation	• Running 2-3 experiments at all times • Quarterly knowledge portfolio reviews • Learning cycles under 30 days
Portfolio	Single Path → Multiple Options	• Asymmetric opportunity seeking • Capability diversification • Strategic relationship building	• Multiple professional trajectories active • High-upside, low-downside projects • Skills that appreciate in emerging domains
Network	Solo Expertise → Collaborative Value	• Knowledge exchange • Connection facilitation • Dynamic reconfiguration	• Active knowledge sharing systems • Cross-domain collaboration • Fluid professional positioning

THE FIVE PRACTICES OF HIGH-VELOCITY ADAPTERS

1. Horizon Scanning

- o Multi-horizon monitoring (0-12 months, 1-3 years, 3-7 years)
  - o Cross-domain exploration
  - o Pattern recognition
2. Learning Acceleration
   - o Meta-learning optimization
   - o Deliberate knowledge experimentation
   - o Knowledge network leveraging
3. Identity Evolution
   - o Identity expansion
   - o Core purpose anchoring
   - o Possible self-exploration
4. Recovery Systems
   - o Failure integration
   - o Energy management
   - o Support network cultivation
5. Opportunity Architecture
   - o Option generation
   - o Strategic positioning
   - o Serendipity amplification

THE THREE CHOICES THAT DEFINE YOUR FUTURE

1. Reposition Your Value Are you operating at a level where AI complements you rather than replaces you?
2. Amplify with AI Are you leveraging AI as a tool to multiply your impact, or are you waiting to be disrupted?
3. Execute for Impact Are you taking action today, testing, experimenting, and learning—or are you waiting for certainty that will never come?

Your Adaptation Velocity Self-Assessment

Rate yourself from 1 (not at all) to 5 (completely) on each item:

Evolutionary Mindset

- I regularly experiment with new approaches before they're required
- I systematically extract learning value from professional setbacks
- I deliberately evaluate which capabilities are appreciating vs. depreciating
- I actively seek "next practices" rather than implementing only established approaches

Portfolio Mindset

- I maintain multiple potential professional paths simultaneously
- I pursue asymmetric opportunities with high upside and limited downside
- I invest in capabilities before their value is widely recognized
- I make decisions based on option creation rather than immediate outcomes

Network Mindset

- I create value primarily through connection rather than solo expertise
- I systematically share knowledge and resources rather than hoarding them

- I reconfigure my professional positioning based on emerging needs
- I develop understanding across domains to facilitate specialist collaboration

Total Score: _____ / 60

75-100%: Extraordinary adaptation velocity

50-74%: Solid adaptation with specific enhancement opportunities

25-49%: Adaptation approach needs significant transformation

<25%: Highly vulnerable to accelerating change

THREE THINGS FOR THIS WEEK

Reality Check: You Don't Need Permission to Reinvent Yourself

1. Reflect: What's your biggest mindset shift since reading this book? Write it down.
2. Draft a one-year AI Adaptation Plan—what's your next big move?
3. Take one bold action this week to AI-proof your career—experiment, launch, or learn.

THE FINAL CALL

LEADING IN THE AI ERA

SARAH'S STORY: REWRITING THE RULES OF AI ADAPTATION

Sarah, a financial analyst, could have panicked when AI took over financial modeling at her firm. Instead, she adapted.

Rather than seeing herself made obsolete, she repositioned her value. She began using AI to generate multiple scenario analyses—work that would have taken her weeks manually—while she focused on identifying which scenarios warranted deeper investigation based on market trends, client relationships, and business intuition that AI couldn't fully grasp.

Sarah didn't just use AI—she directed it, challenged its assumptions, and applied her judgment where it mattered most. Within a year, she wasn't just keeping up—she was promoted to lead a new augmented analysis team, serving twice as many clients with greater strategic insight.

Sarah's story is a blueprint for AI independence. The professionals who thrive in the AI era are not those who resist change, nor those who blindly embrace every new tool. They are the ones who take control—who use AI as a force multiplier while ensuring their skills, their judgment, and their value remain uniquely human.

AI WILL NOT REPLACE YOU. BUT WAITING FOR CERTAINTY WILL.

Your future isn't written yet. You get to define it. AI won't do that for you—but it can amplify everything you choose to create.

AI independence is not about rejecting AI. It is about leveraging it with purpose, ensuring that your decisions, your skills, and your career path remain driven by human intelligence, not artificial convenience. It is about recognizing that AI is not a strategy, but a tool—one that can amplify your impact but should never define your value.

THE FIVE PILLARS OF AI INDEPENDENCE

Reposition Your Value

The most successful professionals understand that their worth is not tied to a specific job description or technical skill set. Instead, they focus on the outcomes they create—the insights they provide, the problems they solve, the innovations they drive.

AI can replicate tasks, but it struggles to replace strategic thinking, contextual awareness, and human judgment. Those who redefine their roles around these strengths will remain indispensable.

Augment, Don't Automate

Technology should expand your capabilities, not replace your thinking.

The professionals who thrive will be those who learn how to direct AI, shaping it to serve their goals rather than passively following its recommendations. They will know when to trust AI and when to challenge it, when to accelerate work with AI, and when to slow down and apply uniquely human discernment.

Build a Resilient Skill Stack

The half-life of skills is shrinking, making continuous learning the only true job security.

The most successful professionals will develop an adaptive mix of technical fluency, creative problem-solving, and cross-disciplinary thinking. They will cultivate curiosity, embrace complexity, and see reinvention as a lifelong process rather than a temporary adjustment.

Control the Narrative

In an AI-driven world, reputation is built not only on expertise but on strategic visibility.

The professionals who stand out will be those who share their insights, document their experiments, and position themselves as forward-thinking leaders. They will not wait for titles or credentials to validate their work—they will demonstrate their value in real time.

Take Action Before You Feel Ready

AI is evolving too fast for anyone to have perfect clarity.

The professionals who thrive will not be those who wait for certainty, but those who take intelligent risks, test new ideas, and adapt as they go. The future will not reward passive observation. It will reward initiative.

The best way to future-proof your career is to build, experiment, and learn—starting now.

Your Competitive Edge in the AI Era

The professionals who thrive in the coming decades will not be those who master a single technology or methodology.

They will be the ones who master the mindset of perpetual transformation.

They will understand that resilience is not about avoiding change but about moving through it with confidence. That uncertainty is not a threat but a signal to grow. That the most valuable skills are not those that can be automated, but those that allow us to think, adapt, and lead in ways no machine can replicate.

This is not about surviving AI. It is about thriving in an era defined by it.

The future belongs to those who step forward, who take ownership of their learning, their evolution, and their impact.

YOU'VE GOT THIS.

The future belongs to those who take charge of it.

You've spent the last 200+ pages challenging old assumptions, sharpening your perspective, and preparing yourself to lead in a world that won't slow down for anyone.

It won't always be easy. Change never is.

But you have everything you need to succeed—the ability to think critically, adapt, and learn faster than the technology around you.

AI isn't your competitor. It's your tool.

And your value isn't something that can be automated—it's something only you can define, refine, and amplify.

Marcus waited for the old world to return; Elena built a new one. Sarah shows us how to lead it.

Sarah sat in the same office where she had once worried about AI taking her job. Now, she led a team where AI wasn't just a tool—it was an extension of their thinking. "We stopped asking, 'What can AI do for us?' and started asking, 'What can we do with AI?'"

That's the shift.

Your AI-proof future isn't about resisting change. It's about harnessing it. This isn't a battle between human and machine—it's a test of who will adapt and who will be left behind.

The world is shifting, but the edge remains human. AI will optimize, automate, and predict—but it will never experience the spark of curiosity, the flash of insight, or the gut feeling that tells you something is about to change.

The people who thrive in this era won't be the ones who cling to outdated playbooks. They won't be the ones waiting for permission. They'll be the ones who sharpen their uniquely human superpowers—curiosity, creativity, judgment, resilience, influence, empathy, and purpose. AI is an extraordinary tool, but these? These are your unfair advantage.

And if you ever wonder whether you're moving fast enough, whether you're making the right move, or whether you're even capable of adapting—pause. Listen. That tingle of an idea? That tiny, unmistakable twitch when something inside you stirs—

curiosity, anticipation, the hunger to explore the unknown? That's your signal.

AI can analyze patterns, but it will never sense that moment. You can. The question is: Will you act on it?

The world is splitting. Those who command AI—and those who are commanded by it. The AI-proof future isn't a theory. It's a choice. You know the rules. Now—how will you rewrite them?

The future won't wait. Neither should you.

APPENDIX 1

The Top Ten Things You Should Get On With
After Reading This Book

You don't need more time. You need a shift.

If you've made it this far, you already know: AI isn't waiting. The rules of work have changed. You can either tinker at the edges—or move with intention and force.

This isn't a summary. It's a hit list. These are the ten things you should actually do—not someday, but now. Not in theory, but in motion.

Start anywhere. Just don't stay still.

1. Run your AI Vulnerability Audit. (Chapter 1)

   Find your exposure before the market does. Be brutally honest.

2. Craft your unique skill stack. (Chapter 7)

   Combine what AI can't do with what only you can do. This is your future résumé.

3. Build your proof-of-work portfolio. (Chapter 3)

   Show, don't tell. Start with one project, one artifact, one outcome you can point to.

4. Stop identifying with tasks—start owning outcomes. (Chapter 5)

   Your title doesn't matter. Your impact does.

5. Turn AI into your sidekick, not your boss. (Chapter 9)

   Pick one tool. Use it this week. Make it work for you.

6. Practice curiosity like it's your most valuable skill. (Chapter 6)

   Because it is. Ask better questions. Explore new ideas. Make time to wonder.

7. Start your Pattern Awareness Journal. (Chapter 5)

   Track what you repeat. Then challenge it, change it, or automate it.

8. Move up the Anti-Ladder. (Chapter 2)

   Identify your current zone—Efficiency? Extension? Expertise? Evolution? Then climb.

9. Pick your next reinvention challenge. (Chapter 11)

   A new skill, a new combo, a new direction. Choose it before the world chooses for you.

10. Shift faster. Don't wait to be ready. (Chapter 11)

    Perfection is a trap. Movement is your edge. Take one bold step this week.

REFERENCES

Acar, S., Burnett, C., & Cabra, J. F. (2019). Ingredients of creativity: Originality and more. Creativity Research Journal, 29(2), 133-144.

Adrià, F., Soler, J., & Adrià, A. (2008). A day at elBulli: An insight into the ideas, methods and creativity of Ferran Adrià. Phaidon Press.

Amabile, T. M. (1996). Creativity in context: Update to the social psychology of creativity. Westview Press.

Andreasen, N. C. (2005). The creating brain: The neuroscience of genius. Dana Press.

Andreessen, M. (2023). The techno-optimist manifesto. Andreessen Horowitz.

Arena, M. J. (2018). Adaptive space: How GM and other companies are positively disrupting themselves and transforming into agile organizations. McGraw-Hill Education.

Autor, D. H. (2021). The work of the future: Building better jobs in an age of intelligent machines. MIT Task Force on the Work of the Future.

Beaty, R. E., Benedek, M., Silvia, P. J., & Schacter, D. L. (2016). Creative cognition and brain network dynamics. Trends in Cognitive Sciences, 20(2), 87-95.

Berlyne, D. E. (1960). Conflict, arousal, and curiosity. McGraw-Hill.

Bersin, J. (2025). HR technology 2025: The definitive guide to the new talent technology landscape. Bersin & Associates.

Boden, M. A. (2004). The creative mind: Myths and mechanisms (2nd ed.). Routledge.

Bone, J., Allen, T., Shepherd, C., & Srinivasan, A. (2024). From credits to capabilities: The changing landscape of education and employment. Harvard Business Review Press.

Bourdieu, P. (1986). The forms of capital. In J. Richardson (Ed.), Handbook of theory and research for the sociology of education (pp. 241-258). Greenwood.

Brown, B. (2021). Atlas of the heart: Mapping meaningful connection and the language of human experience. Random House.

Brynjolfsson, E. (2025, January 28). Billions of dollars are being wasted on AI, Stanford expert says. Axios. https://www.axios.com/2025/01/28/davos-ai-companies-investment-returns

Brynjolfsson, E., Benzell, S. G., & Rock, D. (2023). How AI changes the rules: Economic foundations for artificial intelligence–driven opportunities. MIT Sloan Management Review, 64(2), 32-41.

Brynjolfsson, E., Li, D., & Raymond, L. R. (2023). Generative AI at work. National Bureau of Economic Research Working Paper No. 31161.

Brynjolfsson, E., & McAfee, A. (2022). The AI revolution hasn't happened yet. Harvard Business Review, 100(4), 120-129.

Brynjolfsson, E., Rock, D., & Syverson, C. (2019). Artificial intelligence and the modern productivity paradox: A clash of expectations and statistics. In A. Agrawal, J. Gans, & A. Goldfarb (Eds.), The economics of artificial intelligence: An agenda (pp. 23-57). University of Chicago Press.

Burger, E. B. (2014). The 5 elements of effective thinking. Princeton University Press.

Burning Glass Institute & Strada Education Network. (2024). Beyond the degree: Pathway transparency and the future of hiring. https://www.burning-glass.com/research-reports/beyond-degree-2024/

Bursztynsky, J. (2024, April 17). Banks are already using ChatGPT to replace Wall Street's grunt work. Fortune.

https://fortune.com/2024/04/17/chatgpt-ai-banks-wall-street-junior-analyst-grunt-work/

Burt, R. S. (1992). Structural holes: The social structure of competition. Harvard University Press.

Burt, R. S. (2004). Structural holes and good ideas. American Journal of Sociology, 110(2), 349–399. https://doi.org/10.1086/421787

Carnevale, A. P., & Rose, S. J. (2023). The uncertain pathway from education to employment. Georgetown University Center on Education and the Workforce.

Carstensen, L. L., Isaacowitz, D. M., & Charles, S. T. (2023). Taking time seriously: A theory of socioemotional selectivity. American Psychologist, 78(2), 165-181.

Covey, S. R. (1989). The 7 habits of highly effective people: Powerful lessons in personal change. Free Press.

Davenport, T. H., & Ronanki, R. (2018). Artificial intelligence for the real world. Harvard Business Review, 96(1), 108-116.

Dell'Acqua, F., Tosheva, K., & Tezuka, H. (2023). Networked work: How collaborative practices evolve in digital environments. Organization Science, 34(1), 228-249.

Deloitte Insights. (2021). The skills-based organization: A new operating model for work and the workforce. https://www2.deloitte.com/us/en/insights/topics/talent/skills-based-organizations.html

Deloitte Insights. (2023). 2023 Global human capital trends: New fundamentals for a boundaryless world. https://www2.deloitte.com/us/en/insights/topics/talent/global-human-capital-trends.html

Deloitte Insights, & Fortune. (2023, July). Summer 2023 Fortune/Deloitte CEO Survey. https://www2.deloitte.com/us/en/insights/topics/leadership/ceo-survey.html

Detrich, A. (2015). How creativity happens in the brain. Palgrave Macmillan.

Dietrich, A. (2004). The cognitive neuroscience of creativity. Psychonomic Bulletin & Review, 11(6), 1011-1026.

Duckworth, A. L. (2016). Grit: The power of passion and perseverance. Scribner.

Duhigg, C. (2023). Reinvention: The science of helping your brain adapt to change. Random House.

Dweck, C. S. (2006). Mindset: The new psychology of success. Random House.

Edmonson, A. C. (2019). The fearless organization: Creating psychological safety in the workplace for learning, innovation, and growth. John Wiley & Sons.

Education Data Initiative. (2025). Student loan debt statistics. https://educationdata.org/student-loan-debt-statistics

Eloundou, T., Manning, S., Mishkin, P., & Rock, D. (2023). GPTs are GPTs: An early look at the labor market impact potential of large language models. OpenAI Working Paper.

Ericsson, A., & Pool, R. (2021). Peak: Secrets from the new science of expertise (2nd ed.). Houghton Mifflin Harcourt.

Ericsson, K. A., Krampe, R. T., & Tesch-Römer, C. (1993). The role of deliberate practice in the acquisition of expert performance. Psychological Review, 100(3), 363-406.

Fauconnier, G., & Turner, M. (2002). The way we think: Conceptual blending and the mind's hidden complexities. Basic Books.

Federal Reserve Bank of New York. (2024). The labor market for recent college graduates. https://www.newyorkfed.org/research/college-labor-market/college-labor-market_underemployment_rates.html

Feynman, R. P., Leighton, R. B., & Sands, M. (2013). The Feynman lectures on physics, boxed set: The new millennium edition. Basic Books.

Forbes. (2024). Top employers no longer requiring college degrees. Forbes Careers Report. https://www.forbes.com/sites/careers/2024/01/15/top-employers-no-longer-requiring-college-degrees/

Frankl, V. E. (2006). Man's search for meaning. Beacon Press.

Fricker, M. (2007). Epistemic injustice: Power and the ethics of knowing. Oxford University Press.

Fuentes, A. (2017). The creative spark: How imagination made humans exceptional. Penguin.

Gartner, Inc. (2023, October 11). Gartner says more than 80% of enterprises will have used generative AI APIs or deployed generative AI-enabled applications by 2026. https://www.gartner.com/en/newsroom/press-releases/2023-10-11-gartner-says-more-than-80-percent-of-enterprises-will-have-used-generative-ai-apis-or-deployed-generative-ai-enabled-applications-by-2026

Gino, F. (2018). The business case for curiosity. Harvard Business Review, 96(5), 48-57.

Gioia, D. (2011). The art of the creative process. University of Southern California Press.

Gleick, J. (2011). The information: A history, a theory, a flood. Pantheon Books.

Goldsmith, M. (2023). The earned life: Lose regret, choose fulfillment. Currency.

Goodfellow, I., Bengio, Y., & Courville, A. (2016). Deep learning. MIT Press.

Graham, M. (1991). Blood memory. Doubleday.

Granovetter, M. (1973). The strength of weak ties. American Journal of Sociology, 78(6), 1360-1380. https://doi.org/10.1086/225469

Grow with Google. (2023). Digital skills for everyone: Annual impact report. https://grow.google/impact2023/

Gruber, M. J., Gelman, B. D., & Ranganath, C. (2014). States of curiosity modulate hippocampus-dependent learning via the dopaminergic circuit. Neuron, 84(2), 486-496.

Hanna, M., Zhang, J., & Rodriguez, A. (2023). The future of communities of practice in the AI-enabled workplace. Journal of Knowledge Management, 27(4), 1132-1151.

Harari, Y. N. (2018). 21 lessons for the 21st century. Spiegel & Grau.

Harrison, S. H., Sluss, D. M., & Ashforth, B. E. (2020). Curiosity adapted the cat: The role of trait curiosity in newcomer adaptation. Journal of Applied Psychology, 105(4), 487-501.

Hart, V. (2010). Art: A history of changing style. Phaidon Press.

Hastings, R., & Meyer, E. (2020). No rules rules: Netflix and the culture of reinvention. Penguin Press.

Hatano, G., & Inagaki, K. (1986). Two courses of expertise. In H. Stevenson, H. Azuma, & K. Hakuta (Eds.), Child development and education in Japan (pp. 262-272). W.H. Freeman and Company.

Heifetz, R., Grashow, A., & Linsky, M. (2009). The practice of adaptive leadership: Tools and tactics for changing your organization and the world. Harvard Business Press.

Homans, G. C. (1958). Social behavior as exchange. American Journal of Sociology, 63(6), 597-606. https://doi.org/10.1086/222355

Huang, M. H., Rust, R., & Maksimovic, V. (2022). The feeling economy: Managing in the next generation of artificial intelligence (AI). California Management Review, 61(4), 43-65.

Iacoboni, M. (2023). Mirroring people: The science of empathy and how we connect with others (2nd ed.). Picador.

IBM. (2024). The skill shift: AI and the evolving job market. IBM Institute for Business Value. https://www.ibm.com/thought-leadership/institute-business-value/report/skill-shift-2024

Indeed Hiring Lab. (2024). Job market trends: Skills-based hiring report. https://www.hiringlab.org/research/skills-based-hiring-2024/

Innosight. (2021). Corporate longevity forecast: Creative destruction is accelerating. https://www.innosight.com/insight/creative-destruction/

Isaacson, W. (2011). Steve Jobs. Simon & Schuster.

Isaacson, W. (2014). The innovators: How a group of hackers, geniuses, and geeks created the digital revolution. Simon & Schuster.

Jarrahi, M. H., Newlands, G., Lee, M. K., Wolf, C. T., Kinder, E., & Sutherland, W. (2022). Algorithmic management in a work context. Big Data & Society, 9(2), 1-14.

Johnson, W. (2023). Smart growth: How to grow your people to grow your company (2nd ed.). Harvard Business Review Press.

Kang, M. J., Hsu, M., Krajbich, I. M., Loewenstein, G., McClure, S. M., Wang, J. T., & Camerer, C. F. (2009). The wick in the candle of learning: Epistemic curiosity activates reward circuitry and enhances memory. Psychological Science, 20(8), 963–973. https://doi.org/10.1111/j.1467-9280.2009.02402.x

Kashdan, T. B. (2009). Curious? Discover the missing ingredient to a fulfilling life. HarperCollins.

Kashdan, T. B., & Silvia, P. J. (2009). Curiosity and interest: The benefits of thriving on novelty and challenge. In S. J. Lopez & C. R. Snyder (Eds.), Oxford handbook of positive psychology (pp. 367-374). Oxford University Press.

Katz, L. F., & Krueger, A. B. (2019). The rise and nature of alternative work arrangements in the United States, 1995–2015. ILR Review, 72(2), 382-416.

Kaufman, S. B., & Gregoire, C. (2015). Wired to create: Unraveling the mysteries of the creative mind. Perigee.

Kidder, R. M. (2005). Moral courage: Taking action when your values are put to the test. William Morrow.

Kim, S., Kim, D., & Oh, D. (2022). Human-AI cooperation: User perceptions of AI assistance in creative tasks. In Proceedings of the 2022 CHI Conference on Human Factors in Computing Systems (pp. 1-15).

Lakhani, K. R., & Wolf, R. G. (2005). Why hackers do what they do: Understanding motivation and effort in free/open source software projects. In J. Feller, B. Fitzgerald, S. A. Hissam, & K. R. Lakhani (Eds.), Perspectives on free and open source software (pp. 3-22). MIT Press.

Liedtka, J., & Kaplan, S. (2023). Strategic imagination: How to see the future coming. Harvard Business Review Press.

Lin, N. (2001). Social capital: A theory of social structure and action. Cambridge University Press.

LinkedIn. (2024). Global skills report: Bridging the gap in a changing workplace. https://economicgraph.linkedin.com/global-skills-report

Loewenstein, G. (1994). The psychology of curiosity: A review and reinterpretation. Psychological Bulletin, 116(1), 75-98.

Lyons, D. M., Parker, K. J., Katz, M., & Schatzberg, A. F. (2009). Developmental cascades linking stress inoculation, arousal regulation, and resilience. Frontiers in Behavioral Neuroscience, 3, 32. https://doi.org/10.3389/neuro.08.032.2009

Marcus, G. (2023). Rebooting AI: Building artificial intelligence we can trust (2nd ed.). Pantheon.

Marcus, G., & Davis, E. (2019). Rebooting AI: Building artificial intelligence we can trust. Pantheon.

Marcus, G., & Davis, E. (2023). Artificial general intelligence is not imminent. AI Magazine, 44(1), 85-93.

Mayer, K., Johnson, H., & Ramirez, T. (2025). The future of work acceleration: How organizations are adapting to rapid technological change. Harvard Business Review Press.

McKinney, S. M., Sieniek, M., Godbole, V., Godwin, J., Antropova, N., Ashrafian, H., Back, T., Chesus, M., Corrado, G. S., Darzi, A., Etemadi,

M., Garcia-Vicente, F., Gilbert, F. J., Halling-Brown, M., Hassabis, D., Jansen, S., Karthikesalingam, A., Kelly, C. J., King, D., ... Shetty, S. (2020). International evaluation of an AI system for breast cancer screening. Nature, 577(7788), 89-94.

McKinsey & Company. (2018). Skill shift: Automation and the future of the workforce. https://www.mckinsey.com/featured-insights/future-of-work/skill-shift-automation-and-the-future-of-the-workforce

McKinsey & Company. (2023). The state of AI in 2023: Generative AI's breakout year. https://www.mckinsey.com/business-functions/strategy-and-corporate-finance/our-insights/the-state-of-ai-in-2023-generative-ais-breakout-year

McKinsey Global Institute. (2023). The economic potential of generative AI: The next productivity frontier. McKinsey & Company. https://www.mckinsey.com/capabilities/mckinsey-digital/our-insights/the-economic-potential-of-generative-ai-the-next-productivity-frontier

Mewawalla, P. (2025). AI-augmented leadership: How artificial intelligence is reshaping organizational dynamics. Harvard Business Review Press.

Microsoft Corp., & LinkedIn. (2024). 2024 Work Trend Index: AI at work is here. Now comes the hard part. https://www.microsoft.com/en-us/worklab/work-trend-index/ai-at-work-is-here-now-comes-the-hard-part

Mitchell, M. (2019). Artificial intelligence: A guide for thinking humans. Farrar, Straus and Giroux.

Nagle, F., Li, S., & Zhou, A. (2023). The network effect: Why companies should care about employees' LinkedIn connections. Harvard Business School Working Knowledge. https://hbswk.hbs.edu/item/the-network-effect-why-companies-should-care-about-employees-linkedin-connections

National Center for Education Statistics. (2023). Tuition costs of colleges and universities. https://nces.ed.gov/fastfacts/display.asp?id=76

National Student Clearinghouse Research Center. (2023). Current term enrollment estimates. https://nscresearchcenter.org/current-term-enrollment-estimates/

Nussbaum, M. C. (1995). Poetic justice: The literary imagination and public life. Beacon Press.

Ostaseski, F. (2023). The five invitations: Discovering what death can teach us about living fully. Flatiron Books.

Petrou, P., Demerouti, E., & Schaufeli, W. B. (2018). Crafting the change: The role of employee job crafting behaviors for successful organizational change. Journal of Management, 44(5), 1766-1792.

Putnam, R. D. (2000). Bowling alone: The collapse and revival of American community. Simon & Schuster.

PwC. (2021). The economic impact of artificial intelligence. https://www.pwc.com/gx/en/issues/artificial-intelligence/economic-impact.html

PwC. (2024, December 17). 2025 AI Business Predictions. https://www.pwc.com/us/en/tech-effect/ai-analytics/ai-predictions.html

Rawls, J. (1971). A theory of justice. Harvard University Press.

Riess, H. (2018). The empathy effect: Seven neuroscience-based keys for transforming the way we live, love, work, and connect across differences. Sounds True.

Robson, K. (2025, January 16). Tech giants lay off thousands in 2024 to focus on AI in 2025. CCN.com. https://www.ccn.com/news/technology/biggest-tech-layoffs-in-2024-2025-focus-on-ai/

Runco, M. A. (2014). Creativity: Theories and themes: Research, development, and practice (2nd ed.). Academic Press.

Ryan, R. M., & Deci, E. L. (2000). Self-determination theory and the facilitation of intrinsic motivation, social development, and well-being. American Psychologist, 55(1), 68-78.

Smith, Z. (2009). Changing my mind: Occasional essays. Penguin Books.

Sotomayor, S. (2016). My beloved world. Vintage.

Southwick, S. M., & Charney, D. S. (2018). Resilience: The science of mastering life's greatest challenges (2nd ed.). Cambridge University Press.

Stack Overflow. (2023). 2023 Developer survey: AI tools and workforce impact. Stack Overflow.

Strada Education Network. (2023). The value of education: Perspectives from learners and employers. Strada Education Consumer Insights. https://stradaeducation.org/report/the-value-of-education/

Surden, H. (2019). Artificial intelligence and law: An overview. Georgia State University Law Review, 35(4), 1305-1337.

Surden, H. (2023). The ethics of artificial intelligence in law practice: An analysis. Georgetown Journal of Legal Ethics, 36(1), 45-75.

Taleb, N. N. (2012). Antifragile: Things that gain from disorder. Random House.

Tedeschi, R. G., & Calhoun, L. G. (2004). Posttraumatic growth: Conceptual foundations and empirical evidence. Psychological Inquiry, 15(1), 1-18.

Tegmark, M. (2017). Life 3.0: Being human in the age of artificial intelligence. Alfred A. Knopf.

Thomason, J., & Williams, J. J. (2023). Human-AI collaboration: Principles and challenges. In Proceedings of AAAI-23 Workshop on Human-Centered AI (pp. 78-92).

Thomson Reuters Institute. (2024). 2024 future of professionals report. https://www.thomsonreuters.com/en-us/posts/legal/future-of-professionals-2024/

Uhl-Bien, M. (2007). Complexity leadership theory: Shifting leadership from the industrial age to the knowledge era. The Leadership Quarterly, 18(4), 298-318. https://doi.org/10.1016/j.leaqua.2007.04.002

U.S. Bureau of Labor Statistics. (2022). Number of jobs, labor market experience, and earnings growth: Results from a national longitudinal survey. https://www.bls.gov/news.release/nlsoy.toc.htm

Uzzi, B., Mukherjee, S., Stringer, M., & Jones, B. (2013). Atypical combinations and scientific impact. Science, 342(6157), 468-472.

Wade, N. (2023). The foundation challenges of generative AI. Nature Machine Intelligence, 5(4), 361-365.

Wilson, E. O. (1998). Consilience: The unity of knowledge. Knopf.

Wilson, F. R. (2002). The hand: How its use shapes the brain, language, and human culture. Vintage.

Woolcock, M. (2001). The place of social capital in understanding social and economic outcomes. Canadian Journal of Policy Research, 2(1), 11-17.

World Economic Forum. (2020). The future of jobs report 2020. https://www.weforum.org/reports/the-future-of-jobs-report-2020

World Economic Forum. (2023). Future of jobs report 2023. https://www.weforum.org/reports/the-future-of-jobs-report-2023/

World Economic Forum. (2024). The future of jobs report 2024. World Economic Forum.

World Economic Forum. (2025). Future of jobs report 2025. https://www.weforum.org/reports/the-future-of-jobs-report-2025/

Wrzesniewski, A., & Dutton, J. E. (2001). Crafting a job: Revisioning employees as active crafters of their work. Academy of Management Review, 26(2), 179-201.

Wu, Z., & Dredze, M. (2023). Limitations of large language models in creativity and divergent thinking tasks. In Proceedings of the 61st Annual Meeting of the Association for Computational Linguistics (pp. 4228-4240).

About the Author

David S. Morgan is a bold thinker and seasoned leader at the intersection of technology, innovation, and human potential. With over three decades of experience as an inventor, CEO, and change agent, he's led transformative efforts across sectors—helping people and organizations adapt, grow, and thrive in times of disruption.

In *AI-Proof Manifesto*, David distills his most urgent insights into a practical, punchy roadmap for building a future-ready identity in an AI-driven world. He's known for bridging strategy with soul—bringing clarity, creativity, and conviction to every project.

When he's not writing, speaking, or challenging the status quo, you'll find him exploring the outdoors in New Hampshire.

David welcomes thoughtful dialogue and collaboration. You can reach him at davidceonh@gmail.com.